00-CTS-5

EFFICIENT REDUNDANCY DESIGN PRACTICES

by:

Tony M. Palmer
Maureen C. Ross, P.E.
Instrumentation Testing Association

Bruce Jank, Ph.D., P.Eng.
Canadian Global Environmental Technologies

2003

Water Environment Research Foundation

The Water Environment Research Foundation, a not-for-profit organization, funds and manages water quality research for its subscribers through a diverse public-private partnership between municipal utilities, corporations, academia, industry, and the federal government. WERF subscribers include municipal and regional water and wastewater utilities, industrial corporations, environmental engineering firms, and others that share a commitment to cost-effective water quality solutions. WERF is dedicated to advancing science and technology addressing water quality issues as they impact water resources, the atmosphere, the lands, and quality of life.

For more information, contact:
Water Environment Research Foundation
635 Slaters Lane, Suite 300
Alexandria, VA 22314-1177
Tel: (703) 684-2470
Fax: (703) 299-0742
www.werf.org
werf@werf.org

This report was co-published by the following organizations. For non-subscriber sales information, contact:

Water Environment Federation
601 Wythe Street
Alexandria, VA 22314-1994
Tel: (800) 666-0206
Tel: (703) 684-2452
Fax: (703) 684-2492
www.wef.org
pubs@wef.org

IWA Publishing
Alliance House, 12 Caxton Street
London SW1H 0QS, United Kingdom
Tel: +44 (0) 20 7654 5500
Fax: +44 (0) 20 7654 5555
www.iwapublishing.com
publications@iwap.co.uk

Library of Congress Catalog Card Number: 2003100315
Printed in the United States of America
WERF ISBN 1-893664-70-8 IWAP ISBN: 1-84339-641-6 WEF ISBN: 1-57278-218-8

Instrumentation Testing Association

The research on which this report is based was funded, in part, by the United States Environmental Protection agency through Cooperative Agreement No. CR-827345-04 with the Water Environment Research Foundation (WERF). Unless an EPA logo appears on the cover, this report is a publication of WERF, not EPA. Funds awarded under the Cooperative Agreement cited above were not used for editorial services, reproduction, printing, or distribution.

Water Environment Research Foundation

ACKNOWLEDGMENTS

The Project Team thanks its many contributors who made the valuable information in this report possible, including WERF staff, the project subcommittee, the project advisory panel, surveyed treatment facilities, case study treatment facilities and the U.S. state and Canadian provincial regulators.

Report Preparation

Principal Investigator:
Tony M. Palmer
Instrumentation Testing Association

Project Team:
Maureen C. Ross, P.E.
Instrumentation Testing Association
Bruce Jank, Ph.D., P.Eng.
Canadian Global Environmental Technologies

Project Advisory Panel

Cy Parks
Dow Chemical Company

Charles Ganze, P.E.
Gulf Coast Waste Disposal Authority

Steve Levy, P.E.
Metropolitan Water Reclamation District of Greater Chicago

John Petito, P.E.
New York Department of Environmental Protection

Michael W. Sweeney, Ph.D., P.E.
Louisville and Jefferson County Metropolitan Sewerage District

Zaki Ahmad
City of Houston

William P. Lewis, P.E.
City of Yuba City

William H. Hendrix, Jr., P.E.
Sacramento Regional County Sanitation District

Steve Hallett, P.E.
City of Toledo

Project Subcommittee

James R. Dartez, chair
Royce Technologies

Dean Falkner
Rock River Water Reclamation

Joseph A. Husband, P.E.
Malcolm Pirnie, Inc.

John C. Kabouris, Ph.D., P.E.
Carollo Engineers

Kyriacos Pierides, Ph.D., P.E.
Gannett Fleming

WERF Staff

Research Program Director: Jeff Moeller, P.E.
Project Manager: Roy Ramani, Ph.D.

Case Study Treatment Facilities

Steve Johnsen
Western Lake Superior Sanitary District

Victor Acosta
City of Santa Barbara, El Estero Wastewater Treatment Plant

Bill Hurley
Orange County Utilities, South Water Reclamation Facility

Nitti Surramaniam
City of Toronto, Ashbridges Bay Wastewater Treatment Plant

Surveyed Treatment Facilities

- Anchorage Water and Wastewater Utility
- Central Contra Cost Sanitary District
- City and County of Honolulu
- City of Detroit
- City of Edmonton
- City of Guelph
- City of Indianapolis
- City of Montreal
- City of Owosso
- City of Phoenix
- City of Roseville
- City of Santa Barbara
- City of Stuart

- City of Toronto
- City of Winnipeg
- City of Yuba City
- Columbus Water Works
- County Sanitation Districts of Los Angeles County
- Dow Chemical
- Fairfax County Government
- Greater Vancouver Regional District
- Gulf Coast Waste Disposal Authority
- King County Wastewater Treatment Division
- Little Rock Wastewater
- Madison Metropolitan Sewerage District
- Metro Wastewater Reclamation District
- Metropolitan Council Environmental Services
- Metropolitan Water Reclamation District of Greater Chicago
- Metropolitan St. Louis Sewer District
- Milwaukee Metropolitan Sewerage District
- Northeast Ohio Regional Sewer District
- Orange County Utilities
- Reedy Creek Improvement District
- Sacramento Regional County Sanitation District
- The City of Calgary
- Tulsa Metropolitan Utility Authority
- Western Lake Superior Sanitary District

U.S. State Regulators

- Alabama Department of Environmental Management
- Alaska Department of Environmental Conservation, Southcentral Alaska
- Arizona Department of Environmental Quality, Water Quality Division
- State of Arkansas Department of Environmental Quality
- California State Water Resources Control Board, Division of Clean Water Programs
- Colorado Department of Public Health & Environment, Water Quality Control Division, WQCD-B2
- Connecticut Department of Environmental Protection, Bureau of Water Management
- Delaware Department of Natural Resources and Environmental Control, Division of Water Resources, Surface Water Discharges Section, Discharge Permits Program
- Government of the District of Columbia, Department of Health
- Florida Department of Environmental Protection, Office of Ombudsman, Domestic Wastewater
- Georgia Environmental Protection Division
- State of Hawaii, Department of Land and Natural Resources
- State of Idaho, Department of Environmental Quality

- Illinois Environmental Protection Agency, Permit Section, Division of Water Pollution Control
- Indiana Department of Environmental Management, Office of Water Quality
- Iowa Water Quality Bureau, Department of Natural Resources
- Kansas Department of Health and Environment, Division of Environment, Bureau of Water
- Kentucky Division of Water, Facilities Construction Branch
- Louisiana Department of Environmental Quality
- Maine Department of Environmental Protection
- Maryland Department of the Environment
- Massachuettes Department of Environmental Protection, Bureau of Resource Protection
- Michigan Department of Environmental Quality, Surface Water Quality Division
- Minnesota Pollution Control Agency, Regional Environmental Management Division
- Mississippi Department of Environmental Quality, Office of Pollution Control
- Missouri Department of Natural Resources, Water Issues
- Montana Department of Environmental Quality Community Services Bureau, Permitting and Compliance
- Nebraska Department of Environmental Quality
- State of Nevada Division of Environmental Protection, Bureau of Water Quality Planning
- New Hampshire Department of Environmental Services, Water Division
- New Jersey Department of Environmental Protection, Division of Water Quality
- New Mexico Environment Department, Surface Water Qualtiy
- New York State, Department of Environmental Conservation, Division of Water
- North Carolina Department of Environment and Natural Resources, Division of Water Quality
- North Dakota Department of Environmental Health, Division of Water Quality
- Ohio EPA, Division of Surface Water, Permit to Install and Agriculture Unit
- Oklahoma Department of Environmental Quality, Water Quality Division
- Oregon Department of Environmental Quality
- Pennsylvania DEP, Bureau of Watershed Management, Water Quality Standards
- Rhode Island, Department of Environmental Management, Office of Water Resources
- South Carolina Department of Health and Environmental Control, Bureau of Water
- South Dakota Department of Environment & Natural Resources
- Tennessee Department of Environment and Conservation, Division of Water Pollution Control
- Texas Natural Resource Conservation Commission, Division of Water Quality
- Utah Department of Environmental Quality, Division of Water Quality
- Vermont Department of Environmental Conservation
- Government of the United States Virgin Islands, Department of Public Works
- Virginia DEQ, Water Quality Program
- Washington Department of Ecology, Water Quality Program

- West Virginia DEP, Office of Water Resources
- Wisconsin Department of Natural Resources, Watershed Management, Wastewater Systems Plan Review
- Wyoming Department of Environmental Quality, Water Quality Division, Water and Wastewater Section

Canadian Provincial Regulators

- Alberta Air and Water Branch
- Manitoba Government
- Newfoundland Department of Environment, Water and Wastewater
- Ministère de l'Environnement, Service de l'assainissement des eaux
- Saskatchewan Environment, Environmental Protection Branch

ABSTRACT AND BENEFITS

Abstract:
Equipment redundancy is designed into water and wastewater treatment plants to improve reliability through the provisions of standby equipment or processes to reduce the risk of failure while continuously meeting water quality standards.

Design guidelines are developed to ensure that critical components retain redundant configurations to eliminate system failures while continuously operating the treatment process (e.g., during equipment failures and while performing preventative maintenance that requires equipment to be taken out of service).

The researchers used surveys to identify effective equipment redundancy design practices in the wastewater treatment industry. They also examined the influence of state and federal regulatory requirements on redundant design practices, investigated alternatives to conventional redundancy, and addressed the scope of redundancy design practices of wastewater treatment.

Results outline differences in redundancy requirements that can exist from state to state, identify the most cost-effective options available to address fault-tolerant systems, and provide insight on concepts currently in use that allow maximum facility effectiveness through the use of redundancy alternatives.

This report provides documentation of survey results and applicability of redundancy design practices, explores alternative redundancy design practices through specific case studies, and documents state and provincial regulations and redundancy design guidelines/recommendations for the United States and Canada. It includes recommendations for research of redundancy design practices and discusses the relevancy of providing redundancy design practices to improve treatment facility operational efficiencies and increase facility security.

Benefits:

- Acts as a guidance tool for facilities and regulators to examine redundancy design practices.
- Establishes a definition of redundancy as it applies to the wastewater treatment process.
- Identifies critical redundancy components in the wastewater treatment process.
- Provides insight into successful redundancy design practices.
- Reveals leading methods design and construction to incorporate redundancy.
- Distinguishes the premise of implementation of redundancy design.
- Determines future research needed for redundancy design practices in the wastewater industry.

- Discusses the cost benefits of applying redundancy practices.
- Reports the level of redundancy design required for optimum performance.

Keywords: Redundancy, reliability, regulations, guidelines

TABLE OF CONTENTS

LIST OF TABLES

LIST OF FIGURES

LIST OF ACRONYMS

ADF	Average Daily Flow
BOD	Biochemical Oxygen Demand
CIP	Capital Improvement Project
CWA	Clean Water Act
DAF	Dissolved Air Flotation
DO	Dissolved Oxygen
EPA	Environmental Protection Agency
FOG	Fats, Oils, and Grease
HRT	Hydraulic Retention Times
ID	Identification
IEEE	Institute of Electrical and Electronic Engineers
MCDF	Maximum Capacity Dissolved Flotation
mgd	Million Gallons per Day
MLSS	Mixed Liquor Suspended Solids
NPDES	National Pollutant Discharge Elimination System
OCU	Orange County Utilities
RAS	Raw Activated Sludge
RBCs	Rotating Biological Contractors
RCRA	Resource Conservation and Recovery Act
RCU	Rated Capacity in Use
SSO	Sanitary Sewer Overflow
SWRF	South Water Reclamation Facility
UV	Ultraviolet
WAS	Waste Activated Sludge
WLSSD	Western Lake Superior Sanitary District
WWF	Wet Weather Flow
WWTP	Wastewater Treatment Plant

Executive Summary

ES.1 Background

The wastewater industry began redundancy design practices with fault-tolerant requirements under the directive of water quality regulatory compliance and safety. Results of two U.S. Environmental Protection Agency (EPA) surveys in 1979 and 1980 of over 90 wastewater treatment plants (WWTPs) that researched factors that limit the performance of WWTPs identified treatment plant design inadequacies and their effects on plants. The EPA study determined that seven of the top 10 limiting factors were related to improper design and noted that one of the major problem areas was lack of flexibility of unit processes (Water Pollution Control Federation, 1989). According to the study, the possible effects of inadequate design contribute to the inability to regulate the process control techniques necessary to maintain removal efficiencies. Based on these results, treatment facilities re-evaluated design practices and began to implement equipment redundancy.

ES.2 Project Objectives and Research Approach

This research identifies effective equipment redundancy design practices in the wastewater treatment industry. It also examines the influence of state and federal regulatory requirements on redundant design practices, investigates alternatives to conventional redundancy, and addresses the scope of redundancy design practices of wastewater treatment.

Results outline differences in redundancy requirements that can exist from state to state, identify the most cost-effective options available to address fault-tolerant systems, and provide insight on concepts currently in use that allow maximum facility effectiveness through the use of redundancy alternatives.

The primary objective was to document the best examples of applied conventional and alternative redundancy design practices in municipal and industrial treatment facilities based on site surveys. The results of this investigation reveal:

- Design redundancy practices work
- Why these design practices work

- The premise for these practice
- The level of design redundancy required for optimal performance
- The cost benefits of applying redundancy practices
- How to implement redundancy practices

Case studies show how system redundancy can be effectively used to increase rated plant capacity by incorporating innovative upgrading technologies and advanced automated process control.

A secondary objective was to identify the similarities and differences in redundancy requirements associated with federal and state regulatory agencies as they apply to municipal and industrial wastewater treatment facilities.

The results provide a practical guide for owners, operators, and designers to replicate the documented success for comparable treatment processes regarding equipment redundancy design practices. This guide can be used to make informed decisions regarding redundancy design practices that can:

- Simplify treatment plant design
- Reduce equipment costs
- Reduce energy and chemical costs
- Reduce labor costs associated with operations and maintenance of complex redundant systems
- Increase plant operations efficiency

ES.3 Organization of Report

This report documents survey results, case studies and regulatory interviews developed during this project. Chapter 1.0 provides an introduction to redundancy. Chapter 2.0 introduces treatment facility redundancy design practices survey results including surveyed treatment plants profiles and their general redundancy design practices. Chapters 3.0 through 10.0 provide treatment facility survey results for the following respective treatment processes:

- Collection systems, lift stations, and influent flow (Chapter 3.0)
- Industrial pretreatment (Chapter 4.0)
- Preliminary treatment (Chapter 5.0)
- Primary treatment (Chapter 6.0)
- Secondary treatment (Chapter 7.0)
- Tertiary/advanced treatment (Chapter 8.0)
- Solids processing (Chapter 9.0)
- Disinfection and effluent (Chapter 10.0)

Chapter 11.0 provides an overview of the results from interviews with U.S. state and Canadian provincial regulators regarding redundancy design regulations and guidelines.

Chapter 12.0 provides a the results of case study interviews regarding redundancy design practices with the following treatment facilities:

- City of Santa Barbara, El Estero Wastewater Treatment Plant
- City of Toronto, Ashbridges Bay Treatment Plant
- Orange County Utilities, South Water Reclamation Facility
- Western Lake Superior Sanitary District, Main Treatment Plant

Chapter 13.0 provides a summary of this project's research findings.

ES.4 Results

Equipment redundancy is practiced in the design and operation of water and wastewater treatment plants to improve reliability through the provision of standby equipment or processes that reduce the risk of failure in order to meet water quality regulations or guidelines. Design guidelines are developed to ensure that critical components retain redundant configurations to eliminate system failures while continuously operating the treatment process (e.g., during equipment failure and while performing preventative maintenance that requires equipment to be taken out of service).

For the purposes of this project, redundancy is defined as any of the following units that are not used during normal operations:

- Infrastructure (tanks, etc.)
- Equipment (pumps, motors, etc.)
- Instrumentation (flowmeters, analyzers, etc.)
- Automation and control (computers, power source, PLCs, etc.)

This includes backup, secondary, or items used for more than one purpose (multipurpose units). Redundancy also includes installed units (not spare parts) used for monitoring and control of processes regardless of permit requirements.

Although the definition of redundancy does not include regulatory compliance or reliability standards, the data shows a direct relationship between treatment process reliability, redundancy design, and regulatory compliance. Regulators play a role in ensuring reliability of treatment processes through the approval of treatment plant designs and construction permits as a means of enforcing redundancy design regulations or guidelines.

This regulatory approval process is designed to fulfill the need for the facility to meet their discharge parameters by reliably operating their treatment processes. Thus, reliability of the facility's treatment processes as required by regulatory compliance is a function of redundancy design practices.

Information regarding redundancy design practices in regulations and guidelines was obtained through interviews with U.S. state and Canadian provincial regulators. Researchers contacted all 50 U.S. states and the District of Columbia and received 49 replies. 41%

of U.S. state regulators who responded reported having regulatory requirements for redundancy design practices. In addition, 71% of the interviewed regulators use recommended guidelines or standards for implementing redundancy design criteria. Ten Canadian province regulators provided information. The majority of the Canadian regulators (57%) reported having recommended redundancy design guidelines or standards in place, while 100% noted that local and regional authorities have regulations in place that require redundancy design practices.

The comparison of receiving water discharge parameters with the primary driving force of redundancy design practices revealed that regardless of the type of discharge receiving stream (or discharge permit), all treatment facilities reported operational efficiency and maintenance issues as the major driving force behind redundancy design practices.

Successful implementation of redundancy depends on the identification of critical components in the treatment process. Research data identifies the critical redundancy components by treatment process. The survey data is divided into two major equipment categories:

1. Bypass infrastructure and backup equipment
2. Instrumentation, automation, and control

In addition, research data reveals that the most critical redundant components are

- Pumping to move flows
- Basins and bypasses for hydraulic capacity
- Level sensors, dissolved oxygen (DO) analyzers, and chlorine residual analyzers to detect process changes
- DCS/SCADA systems to monitor, control, and automate the treatment process

Orange County Utilities (OCU) applies redundancy design to their facilities to meet regulatory requirements and to reduce operational costs by reducing process-unit downtime. This level of redundancy is cost effective at OCU because their operating costs have stabilized in the last 3 years, even though reclaimed water production has increased. OCU redundancy levels also resulted in cost savings through reduced onsite staffing needs.

The commitment from Western Lake Superior Sanitary District (WLSSD), staff, management, and owners is reported as one of their most important factors in successful implementation of redundancy design. WLSSD further states that facilities managed with a "performance culture could add redundancy to a continuous improvement plan." The local power company also gave a grant to WLSSD for $141,000 for the net reduction of approximately 700 kW electrical demand due to their redundancy practices.

The City of Toronto stresses that implementing successful redundancy design practices depends on a thorough understanding of the treatment process that require redundant process units, equipment, and automation. The implementation of proven technology is also a large part of their successful redundancy design practices. The City of Toronto has optimized operation and maintenance costs by the use of redundancy and also is able to manage electricity use more effectively.

The City of Santa Barbara states that economics dictate a sufficient level of automation and reliable power sources that established the premise for their redundancy programs. Their redundancy practices have improved their treatment reliability and allowed them to meet National Pollutant Discharge Elimination System discharge standards. They also reported that redundancy requirements have initiated major capital improvement projects that will improve system performance.

ES.5 Recommendations for Future Research

Additional research is needed to address the many variables that can affect redundancy levels at WWTPs. In addition, this project has formulated the many issues that need to be addressed in future planning, design, construction, and regulatory planning of redundancy design practices at WWTPs.

Following is a list of potential strategic research that can be used for identifying future redundancy design research projects:

- Explore the application of redundancy design practices to facilitate emergency, security, and disaster planning issues.
- Research how redundancy is employed during wet weather/peak flow events.
- Using this project's case studies, develop criteria for examination of the impact of redundancy on capital costs, operating and maintenance costs, and system performance.
- Using the revised assessment criteria, increase the number of case studies to obtain a better understanding of the impact of redundancy design on treatment system cost and performance.
- Accomplish documentation of realistic approachs to redundancy guidelines that have practical goals. Various guidelines can be developed in consideration of high-, medium-, and low-cost options, or short- and long-term projects.
- Develop a "how-to implement redundancy design practices" manual using this project's case study treatment facilities as a basis.
- Study redundancy issues facing WWTPs based on the required levels of treatment needed to achieve water quality standards.
- Conduct case study research of the operation of expanded and upgraded WWTPs, taking into consideration the size of each treatment plant, since there are solutions that can be implemented at smaller plants that do not apply to larger systems and vice versa.
- Examine the effective use of redundant instrumentation and automation, which can have a significant impact on design and operation. Study how it can be used to reduce capital costs or eliminate plant expansions and reduce operations and maintenance costs.

- Create a matrix for experimental designs that would provide the option to cover a range of variables, and relate them to the need for redundancy in the design of the treatment plant.
- Conduct further research to review current published redundancy guidelines specifically for redundancy practices used by wastewater treatment facilities and regulators. Redundancy issues for facilities would be addressed based on levels of treatment, and these redundancy levels would be related to the regulatory process.
- Develop a redundancy assessment and design tool to be used by wastewater treatment designers, facility staff, and regulators. This tool would be based on this project's research and successful case studies.
- Create a report that highlights the pros and cons of specific redundancy practices, including recommendations on the best practices.
- Create a document that details how treatment facilities managed to achieve the level of reliability necessary.

CHAPTER 1.0

INTRODUCTION

1.1 Background

Redundant design is applied in many industries that require fault-tolerant systems. Aerospace, nuclear, power, electronics, telecommunications, and computer applications (both hardware and software) industries apply redundant multi-layer design of critical components to ensure fail-safe, dependable, and continuous service. These mostly private industries are typically profit driven, have a diversity of ownership and operating structure, and therefore only use redundancy as absolutely necessary for optimum performance.

Design guidelines for adding redundant or backup components to a system design are developed for industries to ensure that critical components retain redundant configurations in the event of failure (IEEE, 1974-1981). Other key components to redundant design include interchangeable parts so that critical components can be repaired quickly without any interruption in service. Fault-tolerant systems typically use redundancy design to deal with failures associated with wear and manufacturing defects.

Redundancy application is mostly limited to safety-critical and other high-risk, high-budget applications. Some systems are so critical that their design provides full redundancy in every element, to ensure that there is no single point of failure, a key requirement for a truly fault-tolerant system. Upon failure, a critical system component with single, double, or even triple redundancy can continue to operate.

The water and wastewater treatment industry, in comparison to the aforementioned sectors, is largely comprised of publicly owned treatment facilities that have operational priorities that are driven by regulatory requirements to meet water quality standards. The wastewater industry began redundancy design practices with fault-tolerant requirements under the directive of water quality regulatory compliance and safety. Results of two U.S. Environmental Protection Agency (EPA) surveys of over 90 plants in 1979 and 1980 researched the factors that limited the performance of wastewater treatment plants (WWTPs) and identified treatment plant design inadequacies and the effects of these inadequate design practices.

The EPA study determined that seven of the top 10 limiting factors were related to improper design and noted that one of the major problem areas was lack of flexibility of unit processes (Water Pollution Control Federation, 1989). The possible effects of inadequate design contribute to the inability to regulate the process control techniques necessary to maintain removal efficiencies. Based on the results of these earlier studies, treatment facilities re-evaluated design practices and began to implement equipment redundancy.

The justification for including redundant equipment in treatment facilities began with the need to continuously operate treatment processes during equipment failure and while performing preventative maintenance that required equipment to be taken out of service. Redundant design features were added to conventional design practices for wastewater treatment that incorporated duplicate equipment for the "lead-lag" concept to divert flow or perform routine maintenance without incurring treatment process downtime. Implementing redundancy assured that a treatment facility's continuous treatment operations would meet federal and state regulatory requirements that protect the environment and public health and safety.

1.2 Redundancy Design Practice Research Benefits

Identifying effective equipment redundancy design practices for wastewater treatment plants can lead to alternative design practices, such as implementation of automated process control. Alternative redundancy practices can help designers and engineers pursue innovation rather than redundant design and documentation tasks. The procedures and innovative approaches used to upgrade WWTPs are well defined (U.S. EPA, 1989; OMEE, WEAO and EC, 1996; Lidman, 1998). The importance of automation in the upgrading of water and wastewater treatment plants is identified in the proceedings of the 7th International Workshop on Instrumentation, Control and Automation of Water and Wastewater Treatment and Transport Systems (Briggs, 1998). The economic benefits of upgrading wastewater treatment plant are presented as case histories (Daigger and Buttz, 1992) using the process approach developed by Environment Canada's Wastewater Technology Centre (OMEE, WEAO and EC, 1996).

By applying alternative redundancy design practices, WWTPs can optimize treatment processes and thus become more efficient and cost-effective in today's competitive world.

This report emphasizes the importance of adopting new philosophies for project management with holistic approaches to ensure contractor optimization for the use of common equipment, parts and labor to lower overall project costs and everyday operations and maintenance costs.

For example, treatment facilities typically contract design jobs to a consultant and construction work to a contractor. This conventional contracting practice leads to a three-layered communication structure between the owner and the designer, the designer and the contractor, and the contractor and the equipment supplier. Negotiations within this structure, in some cases, can create duplicate designs (i.e., the owner specifies, the designer specifies, and the supplier specifies). All parties involved in the contract have different redundant design priorities and can sometimes specify the same equipment, resulting in excess redundancy. The more recent practice of a competitive design-build contract, design-build-operate, or a design-build-own-operate contract uses one team for the design, construction, and possibly the operation of the plant. The design build practices cannot only reduce project timelines and budgets but can also eliminate "runaway redundancy" design practices.

CHAPTER 2.0

REDUNDANCY DESIGN PRACTICES SURVEY

2.1 Introduction and Background

The main task of this research project included the development of site surveys that were given to wastewater treatment facilities so that researcher could collect redundancy design practice information. The comprehensive survey form (provided in Appendix A) includes a definition of redundancy and outlines redundancy practices in the treatment facility category by infrastructure, equipment, instrumentation, automation, and control. The survey includes the following treatment areas: collection systems; industrial pretreatment; preliminary, primary, secondary, and tertiary/advanced treatment; solids processing; and disinfection and effluent discharge. In addition, the survey collected information regarding the influence or reasons for redundancy or any regulatory guidelines that influence redundancy practices.

A copy of the survey was mailed to municipal and industrial treatment facilities throughout the United States and Canada, primarily to Instrumentation Testing Association (ITA) members and affiliates. A downloadable hard copy version was provided in Adobe Acrobat format on ITA's Web site. The survey was distributed at WEFTEC 2001 in Atlanta, and researchers e-mailed WERF members and requested their participation via the survey on ITA's website.

Forty-six surveys were received from North America, seven from Canada and 39 from the United States, representing 4% (2 facilities) industrial, 7% (3) water, and 89% (41) wastewater treatment facilities.

2.2 Treatment Facility Demographics

The treatment facilities that participated in the redundancy design practices survey are listed in a tabular format in Appendix B, which provides the name and location of these treatment plants. Although this survey reports data using plant identification numbers (Plant ID No.), plant ID numbers are not cross-referenced with the treatment facility names in Appendix B. Due to data entry, not all Plant ID numbers are used (e.g., Plant ID numbers 1 through 3 were tests and are not used in this report).

2.3 General Treatment Plant Profiles

This portion of the survey requested general treatment process information in order to indicate the quantity and quality of plant conditions because these conditions may affect the level of redundancy design practices. General utility and treatment process survey results are summarized below. Graphical displays of these survey results are provided in Appendix C, and individual treatment plant survey responses are provided in a tabular format in Appendix D.

The majority of the plants that participated in this survey treat wastewater at publicly owned and operated facilities with continuous attended operations seven days per week, 24 hours per day. Most of the facilities have a sanitary sewer system only and receive influent flows consisting of municipal combined with pretreated industrial wastewater. Treatment processes are mostly secondary suspended growth and disinfection by chlorination and dechlorination using an automation and control system.

2.3.1 Average Daily Flow And Plant Operations Staff

A comparison of plant average daily flow in million gallons per day (mgd) and the number of total operations staff is shown in Figure 2-1 and 2-2. Figure 2-1 shows all data as reported with a scale of 0 to 850 mgd for average daily flow and total operations staff. Figure 2-1 shows that more than half of facilities reported average daily flows of £100 mgd. Figure 2-2 is rescaled to 0 to 100 mgd for average daily flow and total operations staff to accommodate a more detailed view of the reported data. Approximately 50% of the reporting facilities provide more than one operator per 1 mgd treated.

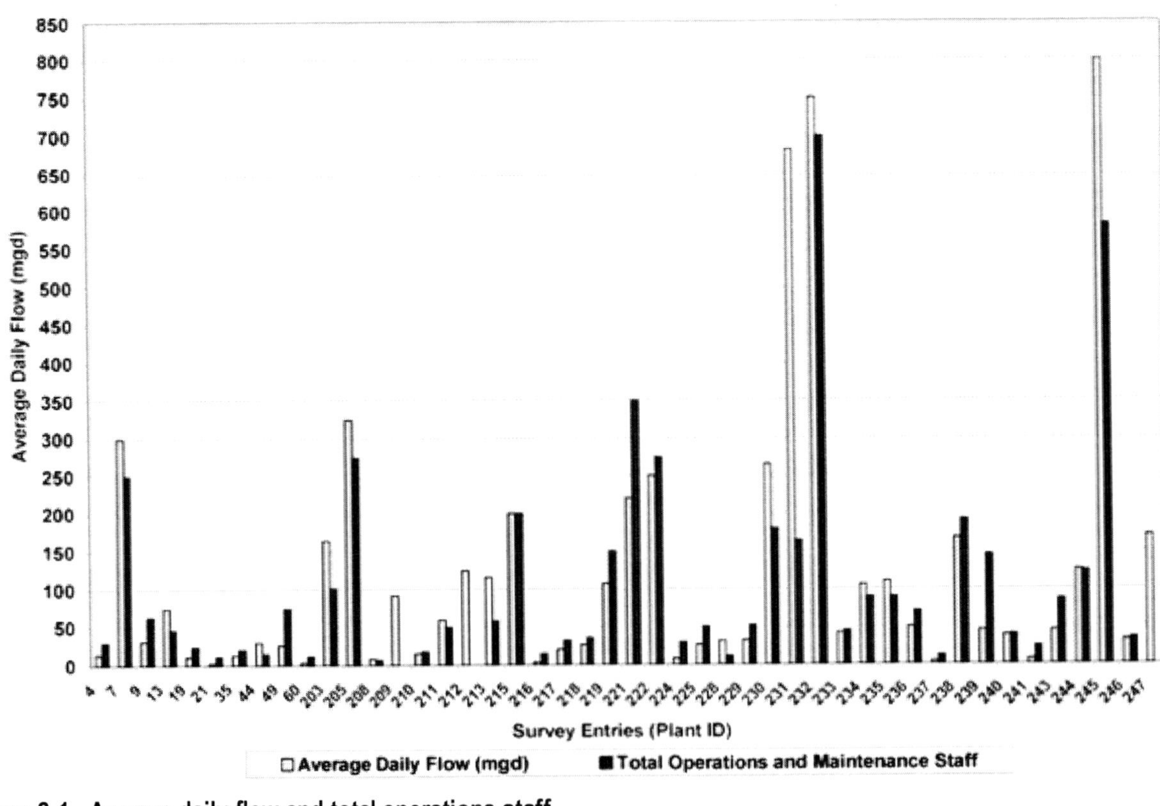

Figure 2-1. Average daily flow and total operations staff.

Water Environment Research Foundation

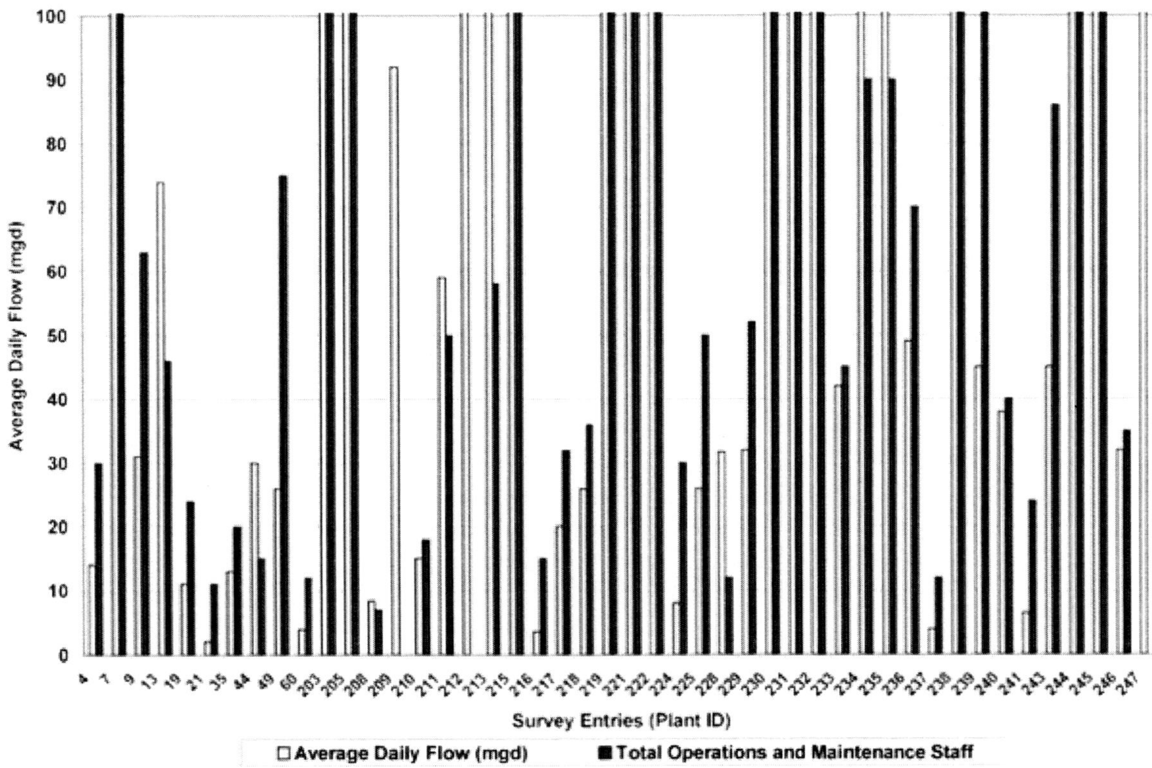

Figure 2-2. Average daily flow below 100 mgd and total operations staff.

2.3.2 Percent Rated Flow Capacity in Use

Facilities reported average daily flow, wet weather flow, and maximum capacity daily flow. Flow data are used to calculate the percent-rated capacity at each facility during average daily flow and wet weather flows. The following equations are used to calculate the percent rated capacity in use:

Percent Rated Capacity in Use (2.3.2-1)
RCU = ADF/MCDF *100

Percent Rated Capacity in Use during Wet Weather Flow (2.3.2-2)
RCUWW = WWF/MCDF*100

where:

ADF	=	Average Daily Flow
MCDF	=	Maximum Capacity Daily Flow
WWF	=	Wet Weather Flow
RCU	=	Rated Capacity in Use

Figure 2-3 was developed using the flow values received from surveyed facilities and eq 2. Figure 2-3 shows that approximately 65% of the facilities operate with less than 50% of rated capacity during average daily flows. During wet weather flows, only 17% of the facilities operate under 50% of rated capacity.

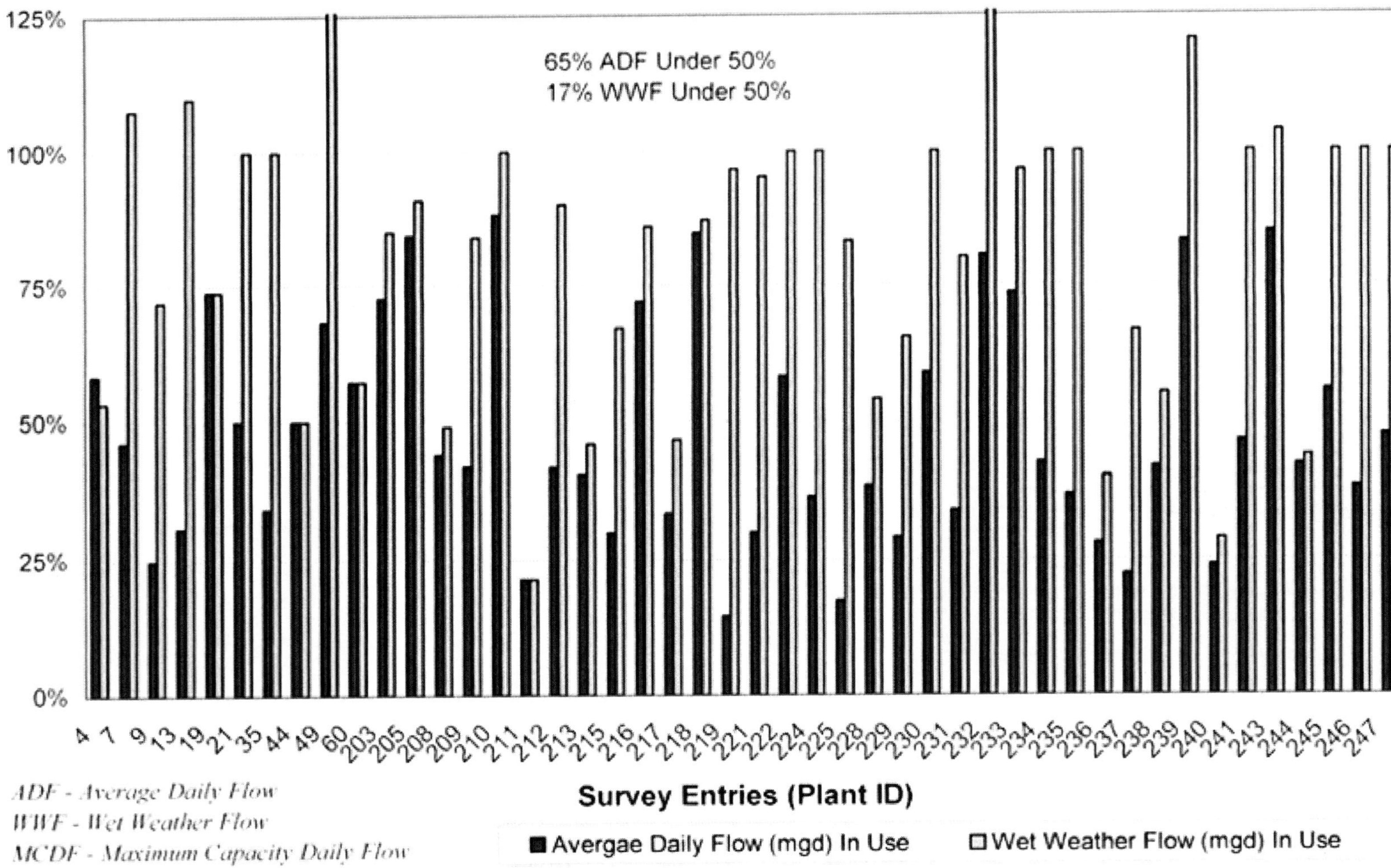

Figure 2-3. Percent rated capacity of design flow in use.

2.3.3 Year Established and Most Recent Upgrade

A comparison of the year the plant was established and the year of the most recent plant upgrade is shown in Figure 2-4. Forty of the 46 plants reported that their most recent upgrade was in the last 10 years.

Figure 2.5 details design and construction methods for the most recent upgrade at reporting plants. Plants marked all methods applicable to their facility. Consultant design is reported for 36% of the plants, and a team approach of in-house staff and consultant design is reported for 24%. 33% of the plants also reported using a construction contractor separate from the consultant. 6% reported using design-bid-build single source. Only 1% of the plants reported using in-house design methods.

Researcher performed an additional analysis (Figure 2-6) to examine the trends of upgrade design methods over the last 30 years. From 1970 to 1980, one plant reported a team approach of in-house staff and consultant design. From 1981 to 1990, one plant reported using in-house and consultant design, whereas four plants reported using a construction contractor separate from a consultant, and four plants reported using an outside consultant design. From 1991 to 2000, 12 plants reported using an outside design consultant, eight reported using the team approach of in-house staff and consultant, 11 reported using construction contractors separate from the consultant, and one plant reported using the design-bid-build method. Finally, from 2001 to 2002, eight plants reported using outside consultant design and construction contractor separate from the consultant, seven plants reported using the team of in-house staff and consultant design, three plants reported using design-bid-build, and one plant reported using in-house design.

2.4 General Redundancy Design Practices

This portion of the survey requested information regarding general redundancy design practices, including redundancy design practices in-place, when these redundancy design practices are used, and the primary driving forces for implementing these practices. All treatment facilities reported having redundancy design practices in place.

Figure 2-7 shows that a majority of the treatment facilities (63%) reported that redundant units are used during peak flows, followed by 29% of the survey respondents who reported that redundant units are not used during peak flows, and 8% who reported that no redundant units are in place.

Receiving water discharge is an important factor in determining permit regulations. The more stringent the effluent quality requirement, the greater the requirement for redundancy in the system design. Figure 2-8 shows that 26% of the surveyed plants discharge into streams designated for recreation, and 22% discharge into other types of streams. 21% of the plants discharge into general use streams, while 12% of the plants discharge into drinking water streams. Ocean discharge accounts for an additional 7% of the reporting facilities, followed by lake and recreation at 5%, and percolation ponds at 2%.

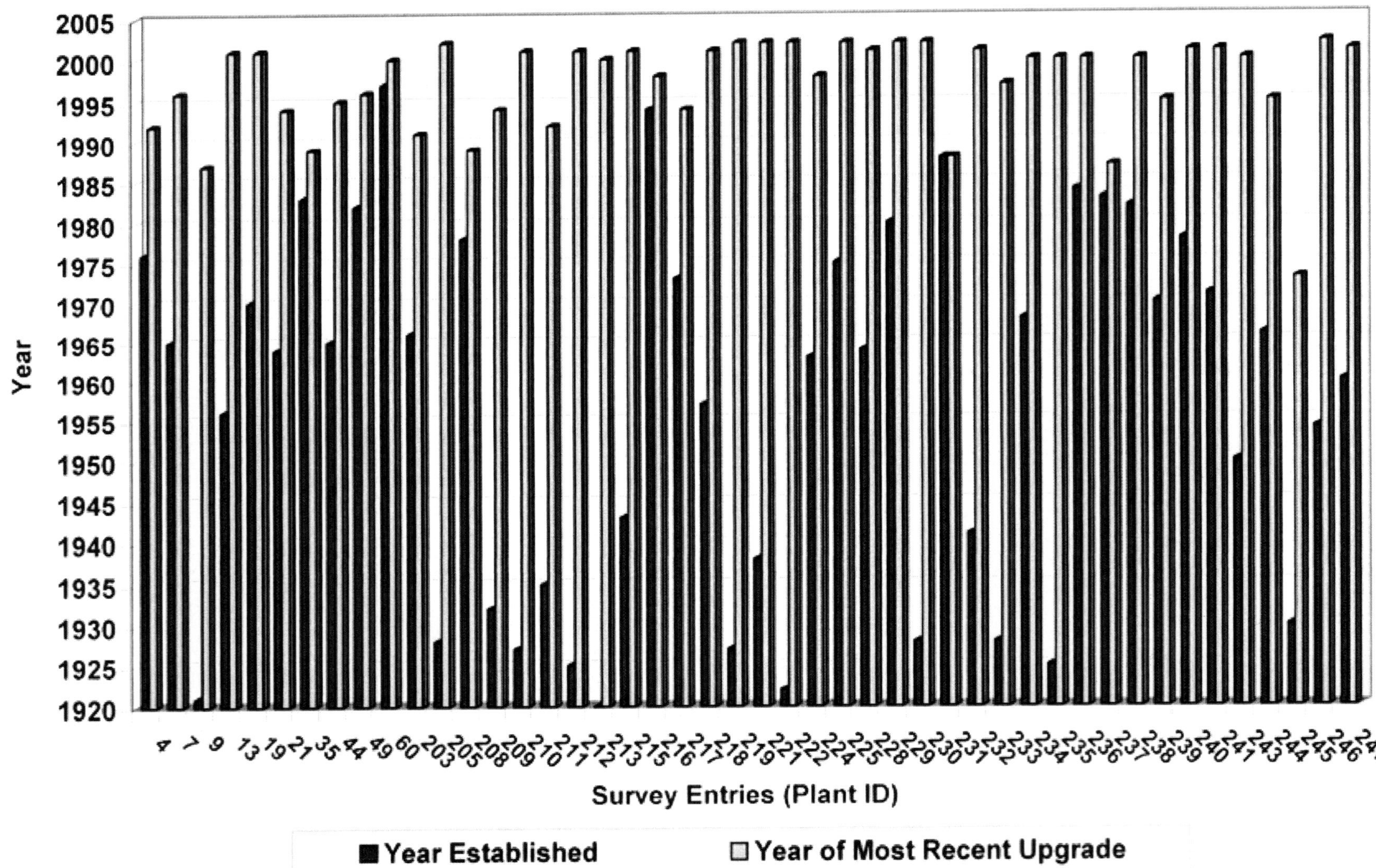

Figure 2-4. Years of plant establishment and most recent upgrade.

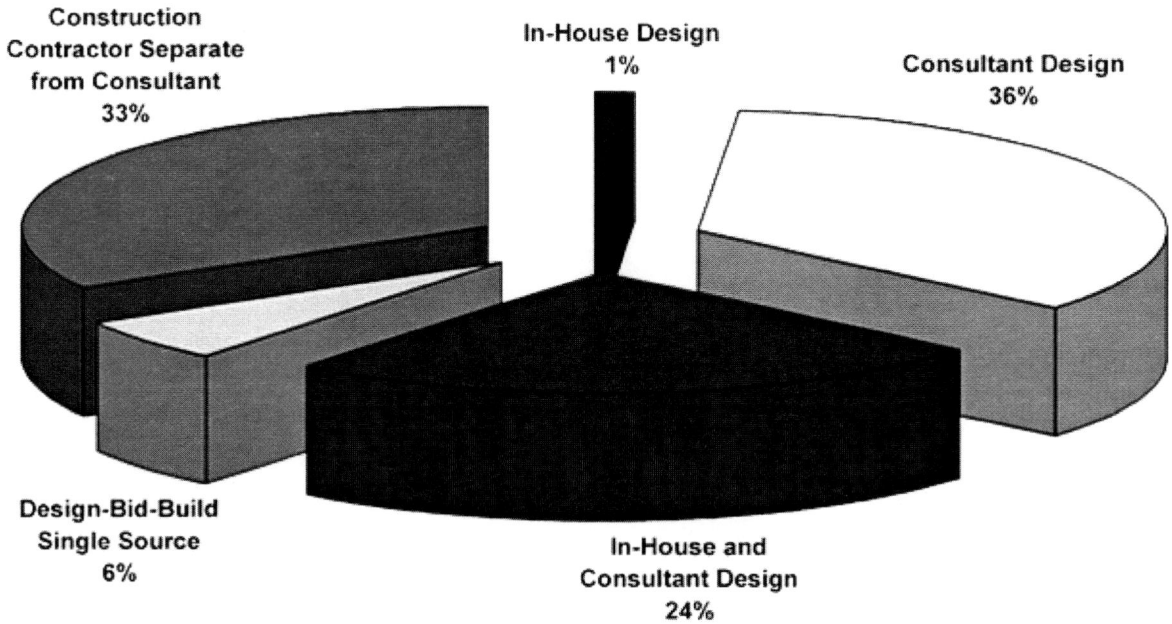

Figure 2-5. Method of design and construction for upgrade.

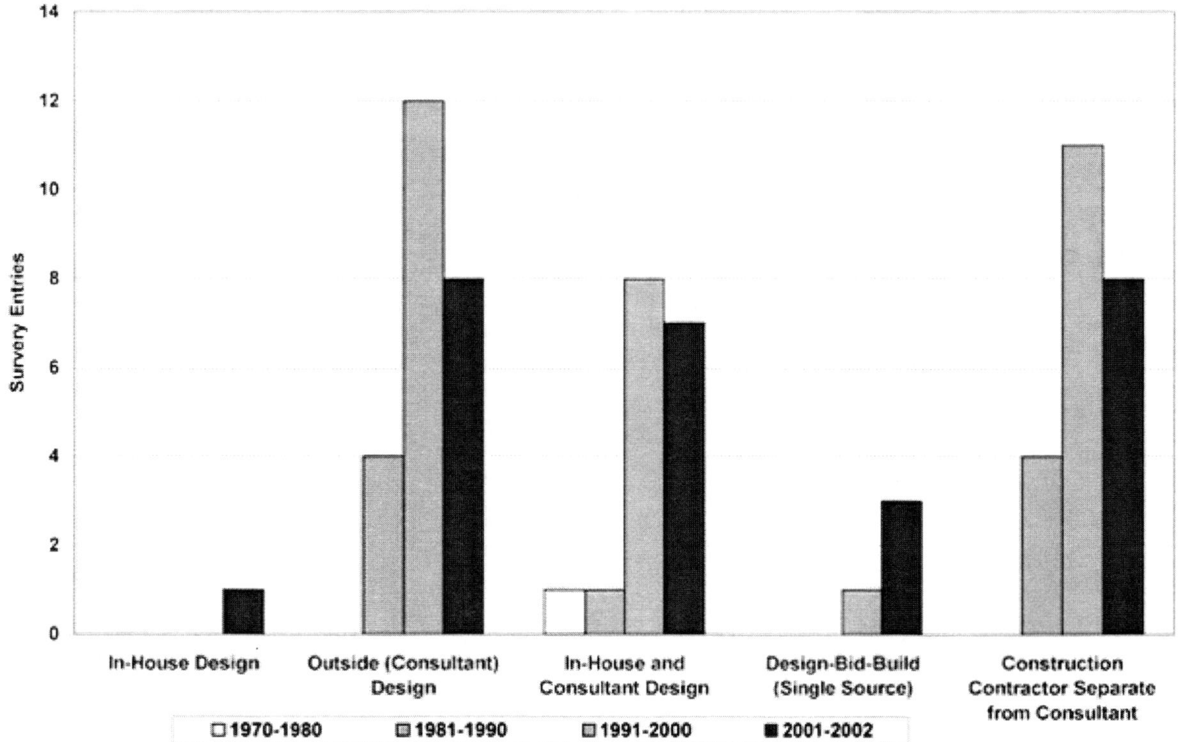

Figure 2-6. Upgrade year vs. upgrade design method.

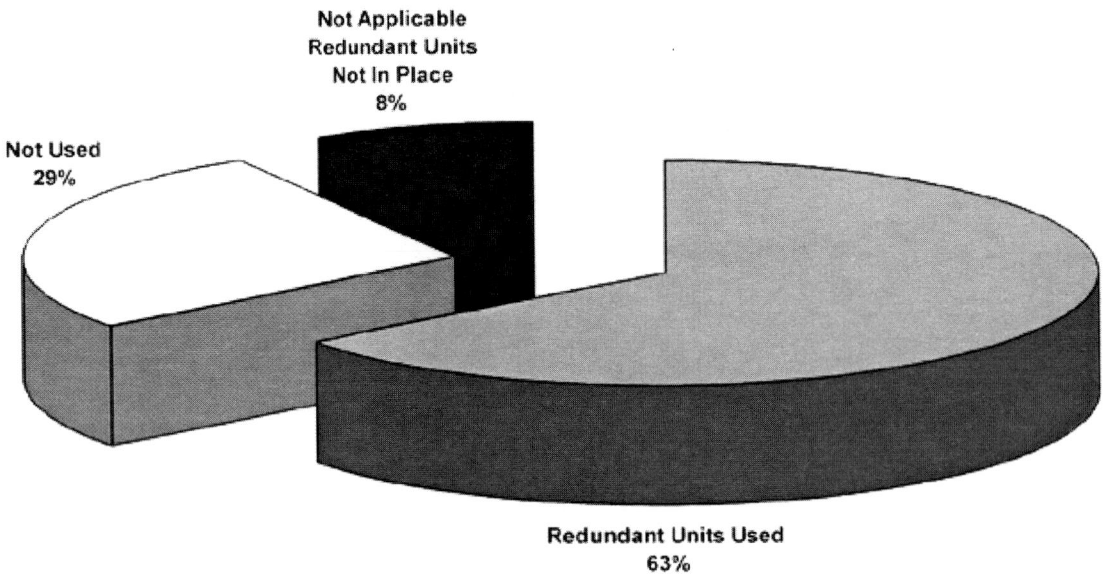

Figure 2-7. Redundant units used during peak flows.

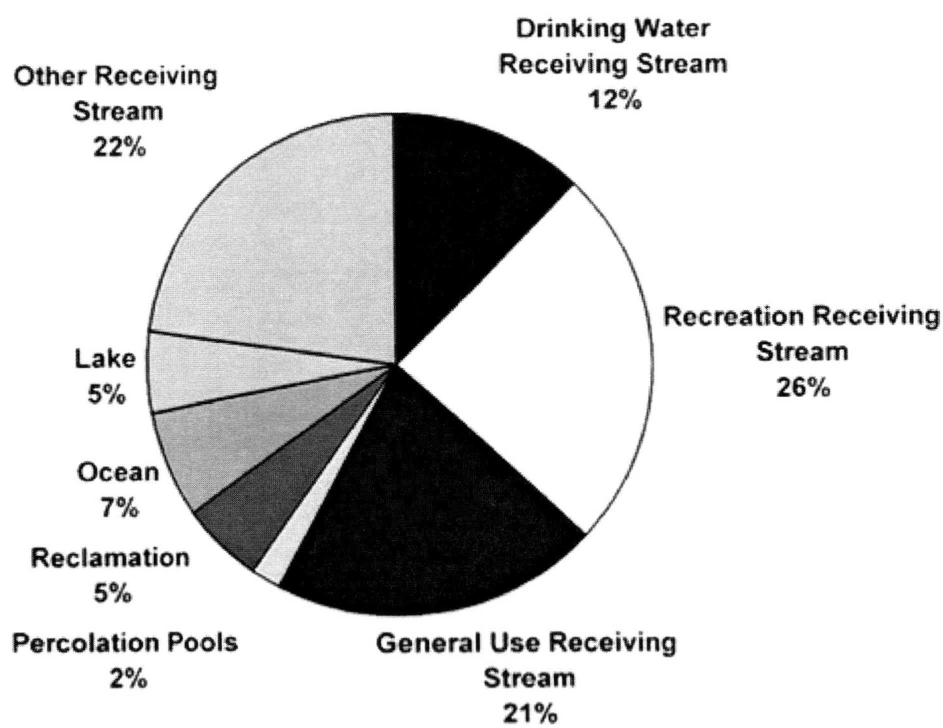

Other Includes: Rapid Infiltration Basins and Reuse; Deep Injection Well; Intermittent Stream; Landscape; River close to Ocean; Navigational water way; Secondary contact; Creek to recreation waters; Dilution > 250 :1; Estuary; Salt Water - Puget Sound; and Seasonal

Figure 2-8. Receiving water discharge.

Figure 2-9 indicates that maintenance issues and operational efficiency were reported by 19% of the facilities as the primary driving force for the origin of redundancy design practices. 14% of the facilities reported state or provincial regulations as the primary driving force for implementing redundancy design practices. Emergency or disaster planning, in-house specifications or design guidelines, federal regulations, and consultant recommendations are the next most frequently reported driving forces, at 12%, 11%, 10%, and 9% respectively. Regional regulations account for 5% of the reporting facilities and equipment supplier recommendations were the primary driving force for 1% of the facilities.

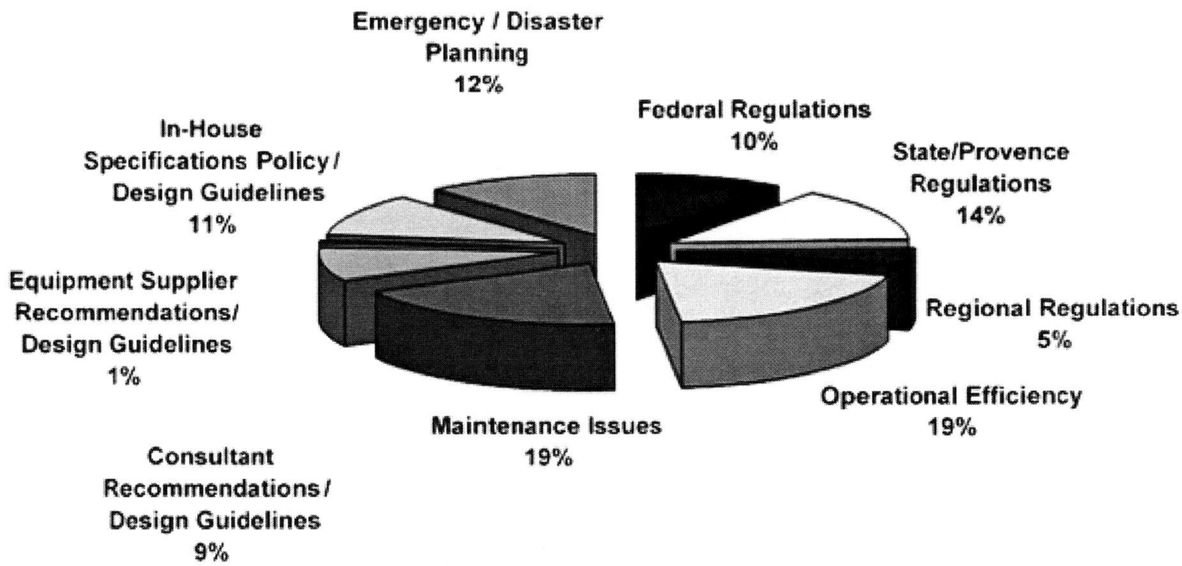

Figure 2-9. Primary driving forces of redundancy design practices.

Figure 2-10 compares reported receiving water discharge parameters with the primary driving force of redundancy design practices. For facilities with drinking water receiving streams, maintenance issues and in-house design guidelines are of primary importance for redundancy design practices, followed closely by operational efficiency.

Facilities that discharge into receiving streams designated for recreation reported in-house design guidelines as the primary driving force, followed closely by operational efficiency and maintenance issues. Facilities with general-use receiving streams reported operational efficiency and maintenance issues were the primary driving force. One facility that discharges into percolation ponds reported emergency/disaster planning as the primary reason for implementing redundancy.

Ocean dischargers listed operational efficiency as the primary reason for having redundancy along with maintenance issues, in-house design guidelines, and emergency/ disaster planning. Lake dischargers reported that regional regulations, operational efficiency, maintenance issues, and consultant recommendations are all of equal importance.

Facilities that discharge into other types of receiving categories list state/provincial regulations as the main driving force behind practicing redundancy. However, operational efficiency and maintenance issues also were important.

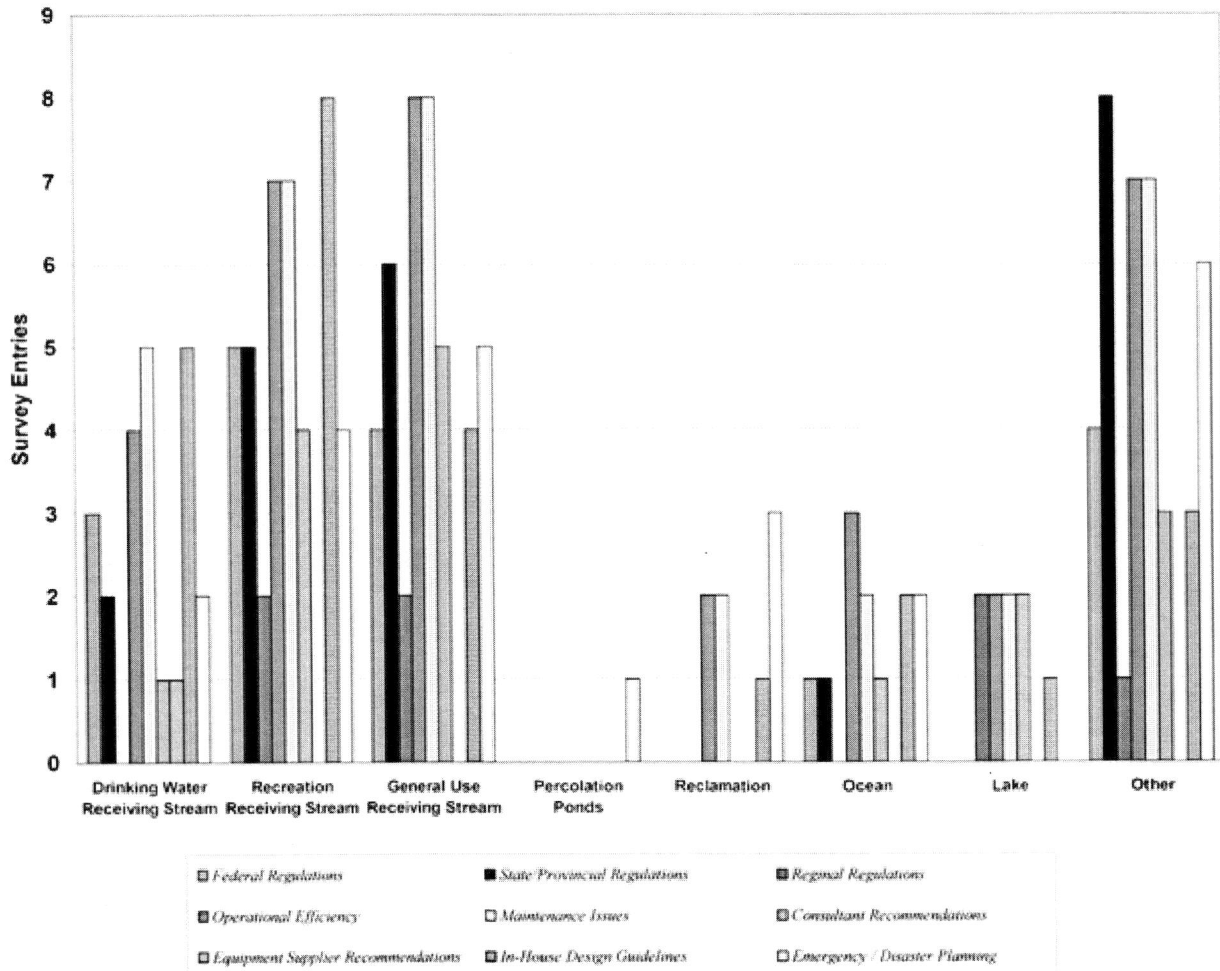

Figure 2-10. Receiving water discharge vs. primary driving force of redundancy design practices.

Other factors also influence the decision-making process for redundancy design. Figure 2-11 shows other means for addressing redundancy by category. The survey indicated that design engineers participate in 90% of the redundancy recommendations, and for 56% of these projects, owners, including in-house engineers and operations and maintenance staff, participate in these recommendations. Supplier recommendations are used by 5% of the facilities for redundancy decisions, and 5% of the facilities reported other reasons for redundancy practices.

Documentation is essential in providing guidelines and justification for redundancy design practices. Figure 2-12 illustrates the type of documentation used by treatment facilities for redundancy practices. Most of the survey respondents (45%) reported that they use operations and maintenance manuals for documenting redundancy, and 30% reported using design report documentation. Approximately 12% reported having no documentation for redundancy practices, and 8% reported documentation of redundancy practices in design manuals. Only 5% reported other means of redundancy documentation, which included treatment facility policy and procedures, historical practices, standard operating procedures, and a reliability standard as two times capacity plus a standby.

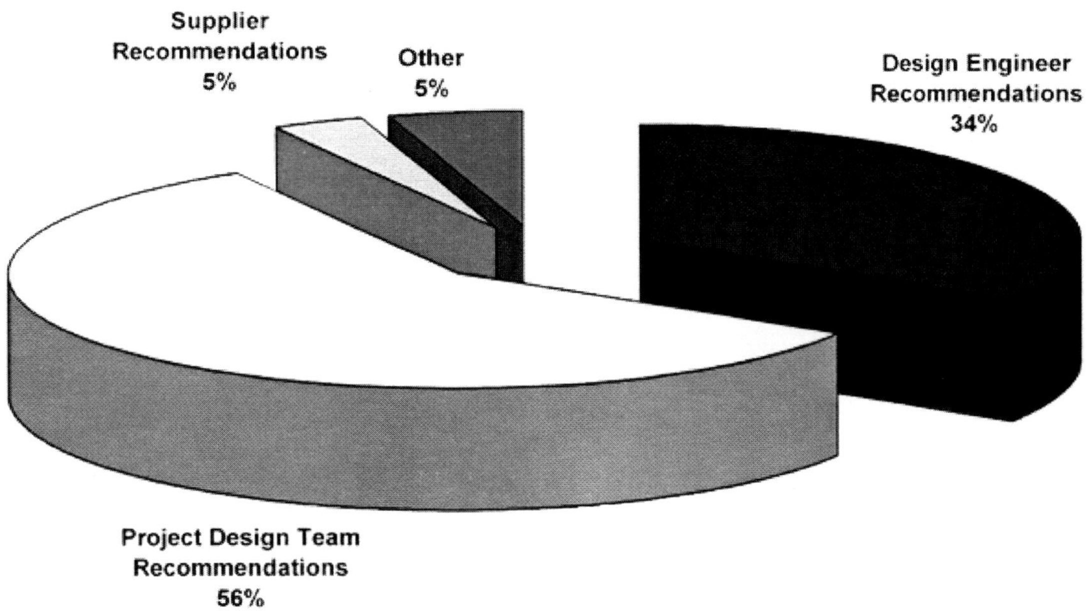

Other Includes: Operations Recommendations; SD 709; No design redundancy program in place

Figure 2-11. Other means of addressing redundancy.

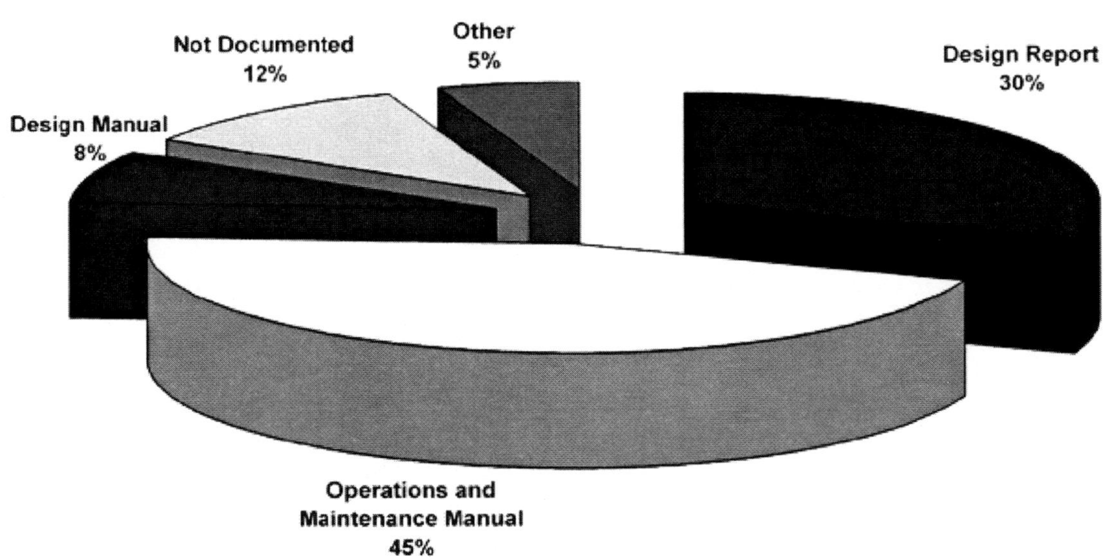

Other Includes: Policy and procedures; Historical Practices; Standard operating procedures; Reliability Standard - two times capacity plus a standby unit

Figure 2-12. Documentation of redundancy design practices.

CHAPTER 3.0

Collection Systems, Lift Stations, and Influent Flow

3.1 Introduction and Background

Collection systems, lift stations, and influent flow works are subjected to severe process conditions caused by weather and industrial and municipal waste flows. Redundancy is a key issue with utilities for these types of facilities because of public health and safety concerns. Sufficient infrastructure and equipment must be built to accommodate future growth with maintenance requirements that lack the need for lengthy shutdowns. Appendix E provides detailed survey results in a tabular format for the collection systems, lift stations and influent flow, and results are summarized below.

3.2 Infrastructure and Equipment

Respondents listed types of bypass units and/or backup, secondary, or multipurpose units in use. Figure 3-1 compiles the responses for infrastructure bypass units of collection systems, lift stations, and influent flow.

Bypass discharge to a local waterbody accounted for 27%, while 20% reported they have bypass influent line or channel into the plant and/or use bypass lines at lift stations. Utilities also reported that 26% have no type of bypass infrastructure for these systems and works. Other novel approaches account for 7%, which includes routing collection system flows to underground storage facilities, pump stations that are not overloaded, and siphon barrel systems.

For backup/secondary infrastructure of collection systems, lift stations, and influent flow, Figure 3-2 shows that 46% of facilities reported having no available backup units. Approximately 20% of the survey respondents reported using influent storage basins or tanks for backup or multipurpose units, and 12% reported using collection systems basins and tanks. One utility reported using the in-line storage capacity of their collection system.

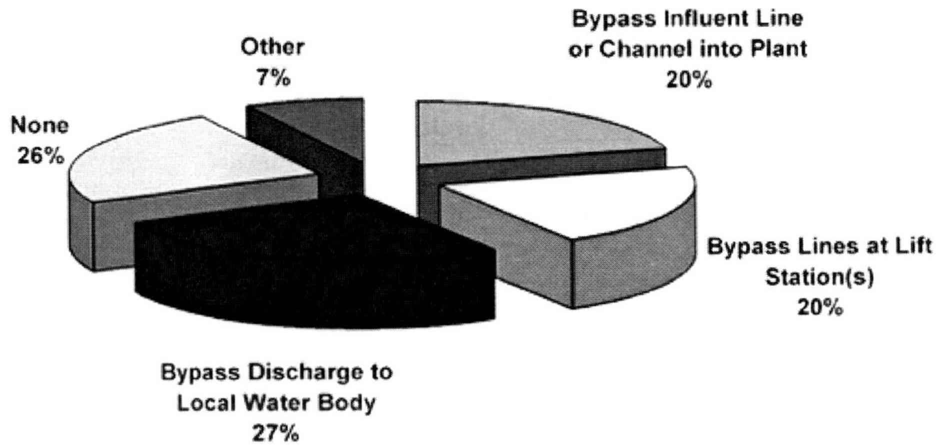

Other Includes: siphon barrel; bypass into TARP system; and ability to route flows from an overloaded pump station to a pump station that is not overloaded.

Figure 3-1. Infrastructure bypass.

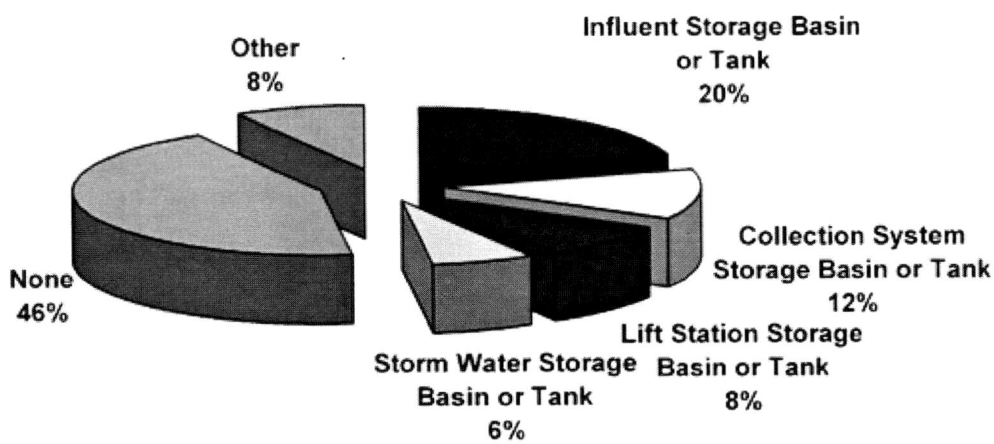

Other Includes: storage in sewerage (in-line); inflatable weir sluice gate; deep tunnel project underway; and tunnel and reservoir plan (TARP).

Figure 3-2. Infrastructure backup units.

Figure 3-3 shows that more than half of the surveyed treatment facilities (58%) reported having redundant pumps at lift stations, 36% reported that portable pumps are used for redundancy, and 8% reported using other methods.

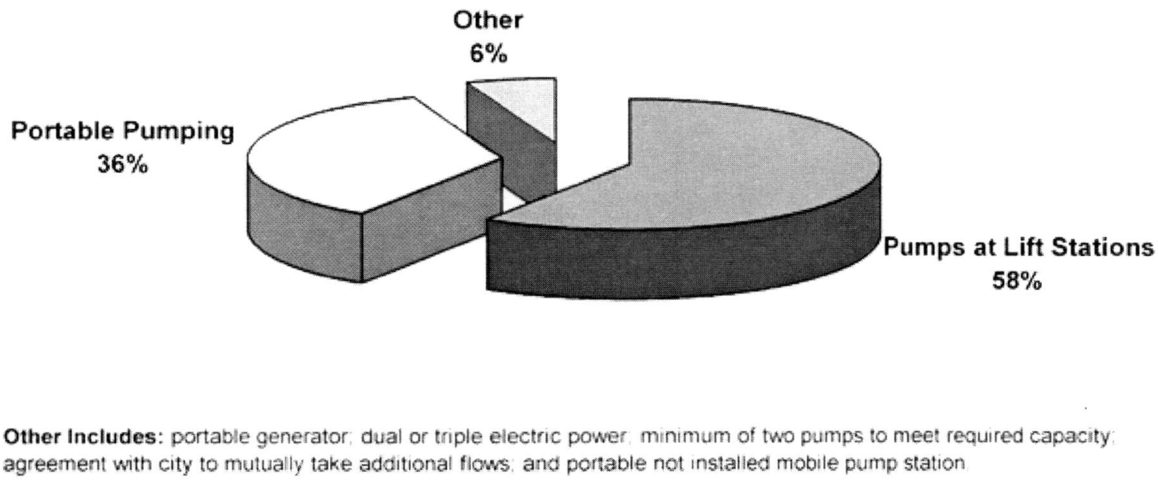

Figure 3-3. Equipment.

3.3 Instrumentation and Automation

Redundant instrumentation of the collection systems, lift stations, and influent flow processes (Figure 3-4), includes backup wet well level sensors as the largest instrument reported at 37%, followed by wet well gas detectors at 12%, and influent flowmeters and collection systems flowmeters both at 8%. 26% of respondents reported that they have no redundant instrumentation for the collection systems, lift stations, and influent flow systems.

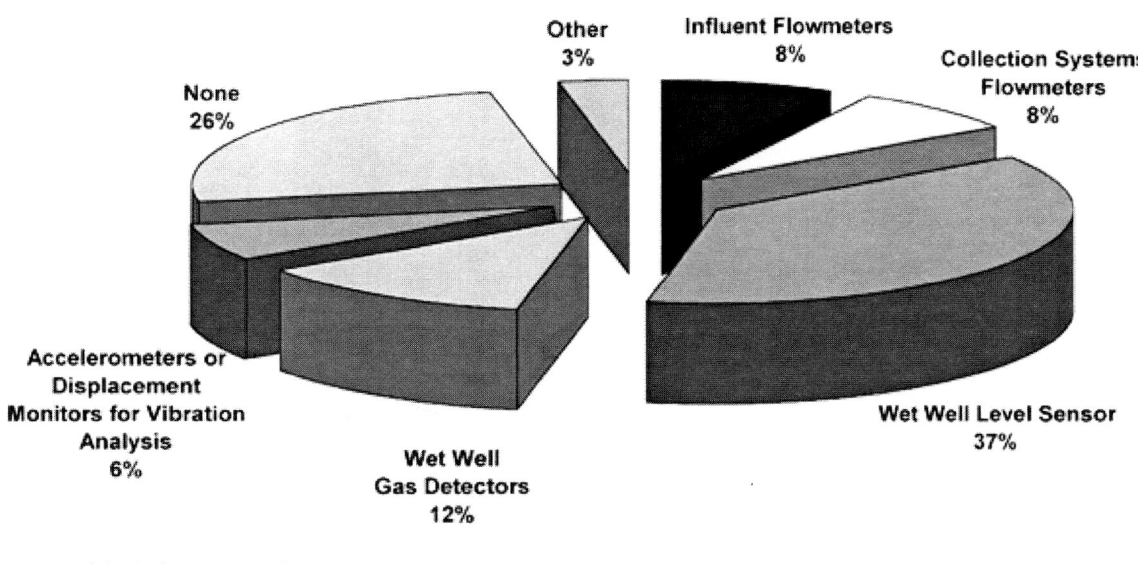

Figure 3-4. Instrumentation.

For automation of the influent flow processes, local and remote control can provide flexibility in process control by allowing treatment plant operators full process control of redundant equipment. Backup power sources and redundant control equipment also are critical.

Figure 3-5 shows that dual power sources are used by 26% of the surveyed facilities and 25% report using portable or in-place generators for collection systems, lift stations, and influent flow. For automation categories, most survey respondents (19%) reported having redundant backup DCS/SCADA systems and (15%) multiple control strategies, while only 7% reported having dual PLCs.

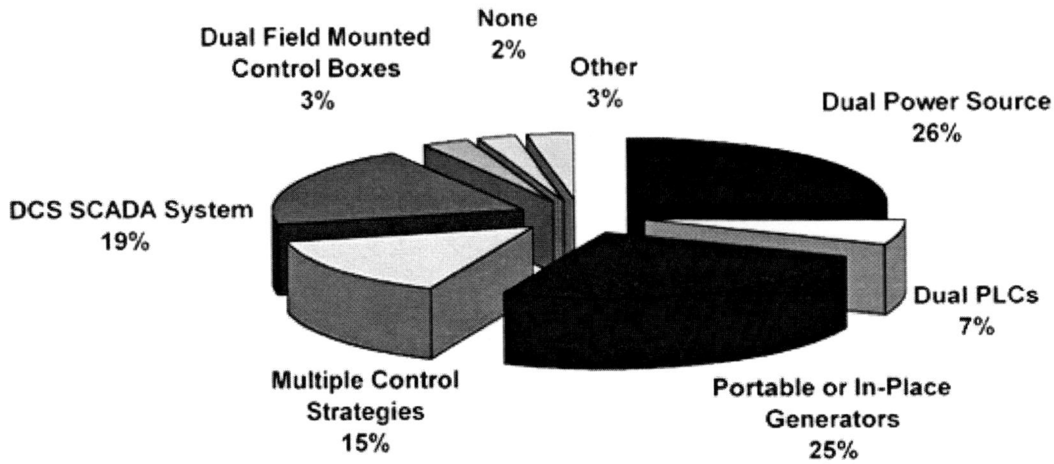

Other Includes: UPS, Alarm RTU (remote terminal unit) via phone line, and 10-States Standard

Figure 3-5. Automation and control.

3.4 Redundancy Reasons and Ratings

Collection systems, lift stations, and influent flow systems can have specific regulations for sanitary sewer overflow (SSO), stormwater, and flood control that can drive a treatment facility's redundancy design practices.

The reported reasons for redundancy are shown in Figure 3-6. Maintenance and operations requirements are reported by 31% for redundant design practices while 20% report emergency/disaster planning requirements. Fifteen percent of the utilities reported sanitary sewer overflow (SSO) regulations, and 10% reported safety regulations. Other utilities reported redundancy design practices influenced by the Water Resource Act, Resource Conservation and Recovery Act (RCRA), and the 10-State Standards.

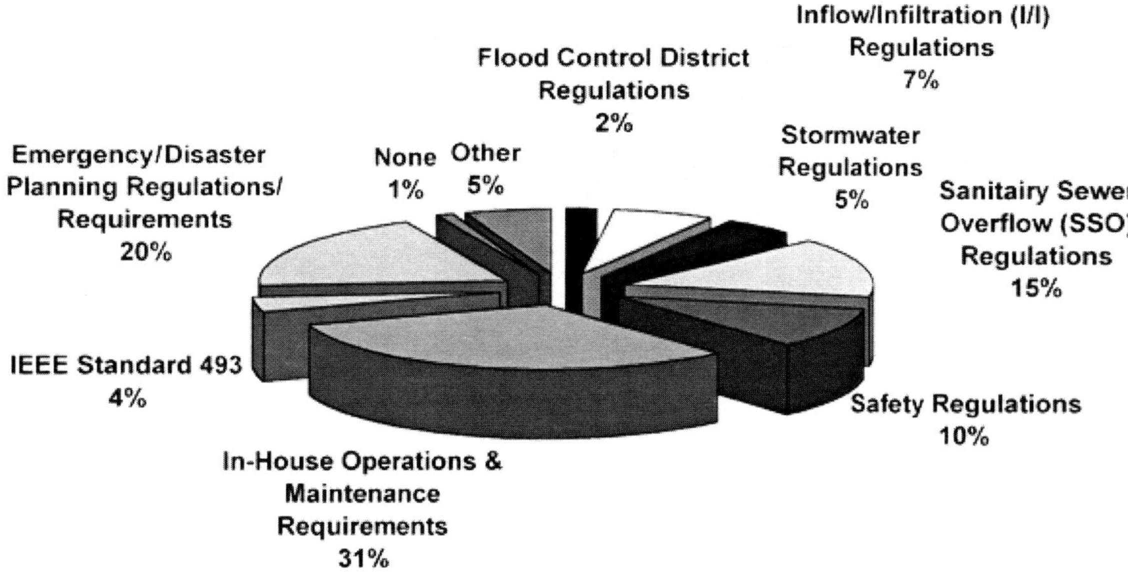

Other Includes: Water Resources Act; RCRA; 10-States Standard; Federal, State and 10 State Standards for plant power redundancy for influent pumping.

Figure 3-6. Reasons for redundancy.

Of the 10 possible reasons for redundancy design practices shown in Figure 3-6, utilities responded to the effectiveness of these practices as shown in Figure 3-7. The majority of the survey respondents (57%) reported that their redundancy design practices are extremely effective. Approximately 38% of the surveyed treatment facilities reported that their redundancy design practices were somewhat effective, whereas, only 5% reported that their treatment plant redundancy for collection systems, lift stations, and influent flow was not necessary. Other treatment facilities reported that their "SCADA and monitoring and backup generators prevented spills and upsets," that "redundancy is used for mechanical failures only," and that the "tunnel and reservoir plan is extremely effective for handling stormwater flow bypass and storage."

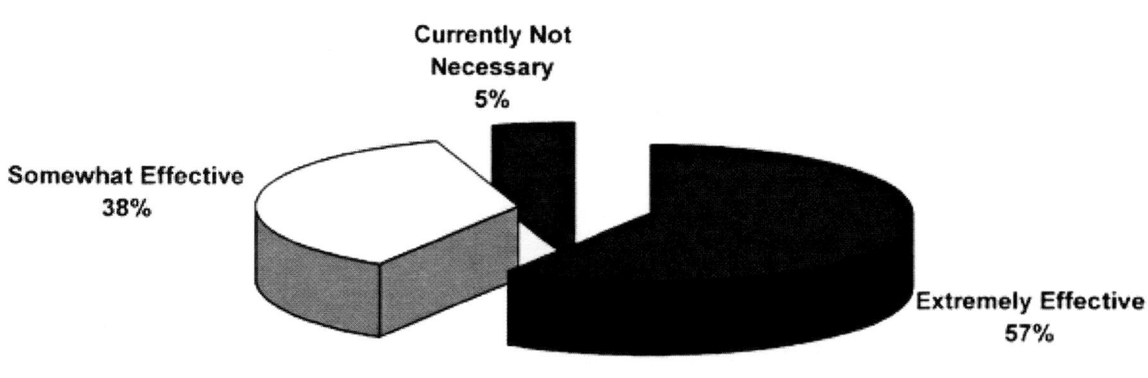

Comments Include: SCADA monitoring and backup generators prevents sewage spills and upsets; The above responses pertain to treatment facility only; No formal redundancy practices in place; A few extremely effective; Not applicable; None; Redundancy used only due to mechanical failure or extended power outage; and TARP (Tunnel and Reservoir Plan) is extremely effective for stormwater bypassing and storage.

Figure 3-7. Redundancy ratings.

CHAPTER 4.0

Industrial Pretreatment

4.1 Introduction and Background

With the advent of General Pretreatment Regulations (U.S. EPA 40 CFR Part 403) in 1978 and the EPA's subsequent changes to these regulations in 1990 (55 FR 30082), industrial utilities must meet specific requirements before their industrial flows are discharged to local wastewater treatment plants WWTPs. These discharge flows may contain high levels of bio-chemical oxygen demand (BOD); metals; organic carbons; fats, oils, and grease (FOG), and other substances, some of which may be toxic.

The management strategies for pre-treating these industrial flows are:

- Identification of wastes generated,
- recycle and reuse of waste streams,
- establishment of pretreatment alternatives, and
- cost studies of those pretreatment alternatives.

WWTPs must then be prepared to process these industrial waste flows reliably and consistently while meeting their own permitting regulatory requirements.

The industrial pretreatment section the survey collected municipal and industrial pretreatment facility information. Appendix F provides detailed survey results in a tabular format for industrial pretreatment, and results are summarized below.

4.2 Infrastructure and Equipment

The basic design of industrial pretreatment includes flow equalization, physical separation (settling), and chemical and biological treatment. Industrial pretreatment infrastructure bypass redundancy units are depicted in Figure 4-1. Flow diversion redundancy for pretreatment accounts for 19% of the reported facilities. This is significant because flow diversion is an integral part of the pretreatment process that allows utilities to better manage varying waste streams and their potential effects on downstream processes. Influent flow channels, gravity separators, and dissolved air flotation (DAF) units are also in use as redundant configurations, as reported by 7% of the survey respondents.

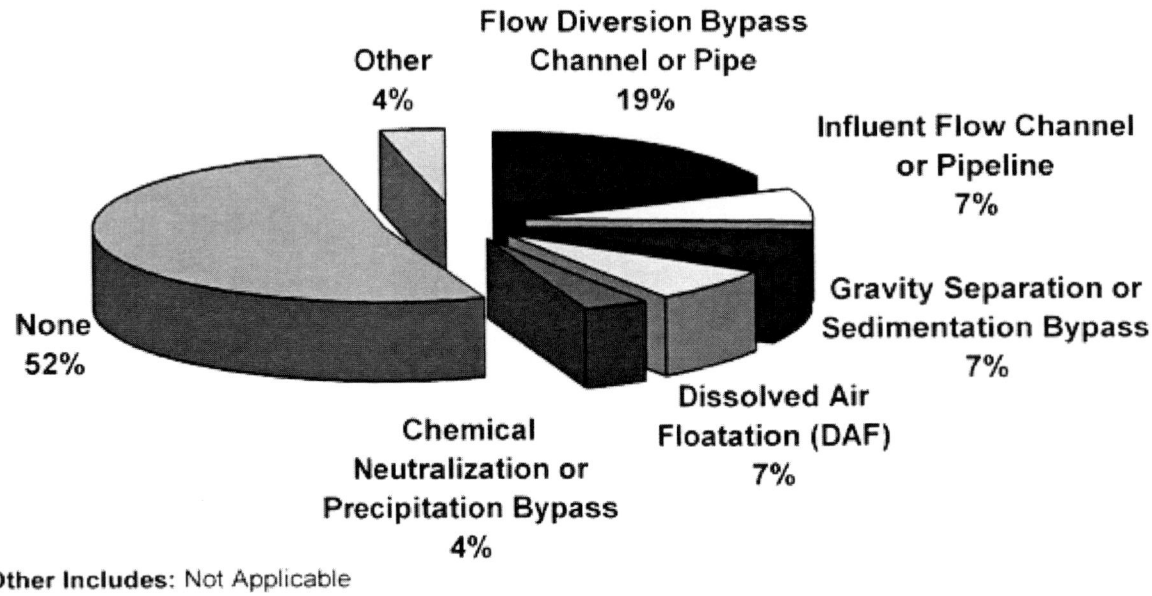

Figure 4-1. Infrastructure bypass units.

Flow equalization backup units are equally important as flow diversion in the management of the pretreatment process. Figure 4-2 shows that surveyed facilities reported that 17% have flow equalization basins for redundancy in controlling pretreatment flows. In addition, 8% reported having redundant gravity separation units and chemical feed tanks. Treatment facilities that do not having an industrial pretreatment process account for the 63% that reported no backup/secondary or multipurpose units for their industrial pretreatment infrastructure.

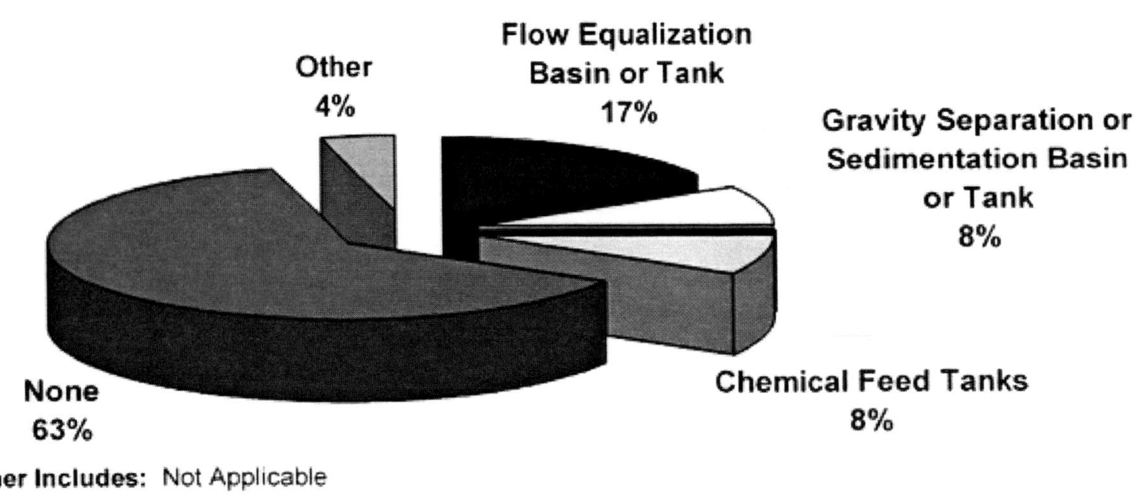

Figure 4-2. Infrastructure backup units.

Figure 4-3 depicts mechanical equipment designated as backup for the pretreatment process. 20% of facilities report that influent pumps are most critical to their industrial pretreatment process, following by sludge pumps, chemical feed pumps, and screens, reported by 16% each. 10% of facilities reported using backup grinders.

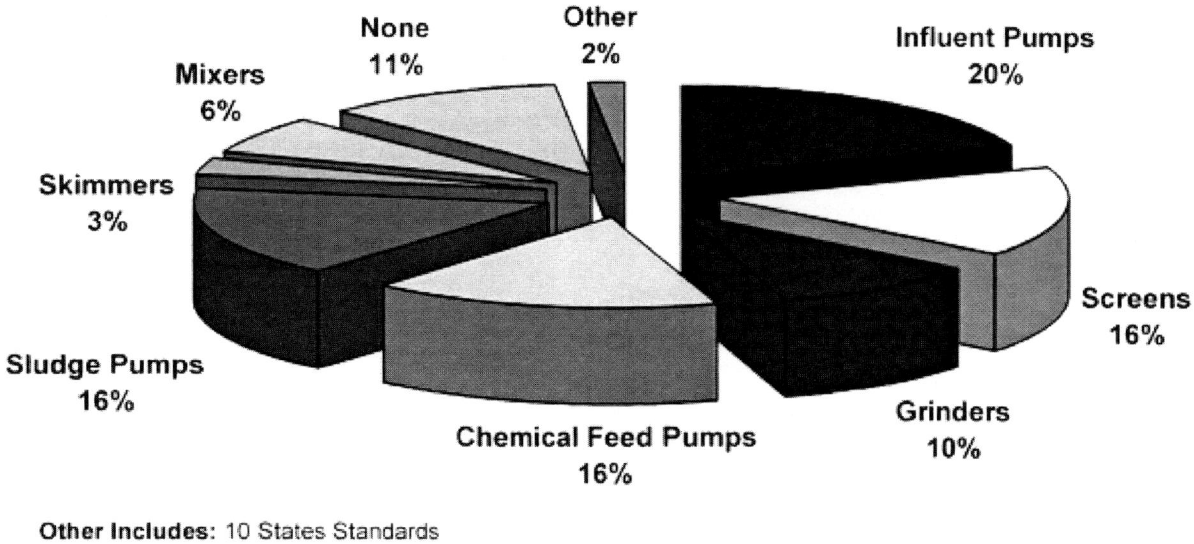

Figure 4-3. Equipment.

4.3 Instrumentation and Automation

The industrial pretreatment process relies on instrumentation and automation to monitor and control flow, settling, and chemical and biological processes. Figure 4-4 shows that backup pH analyzers were reported by 16% of the facilities, and 13% reported having backup flowmeters. In addition, analytical chemistry analyzers and level sensors were reported by 10% of the facilities that practice industrial pretreatment.

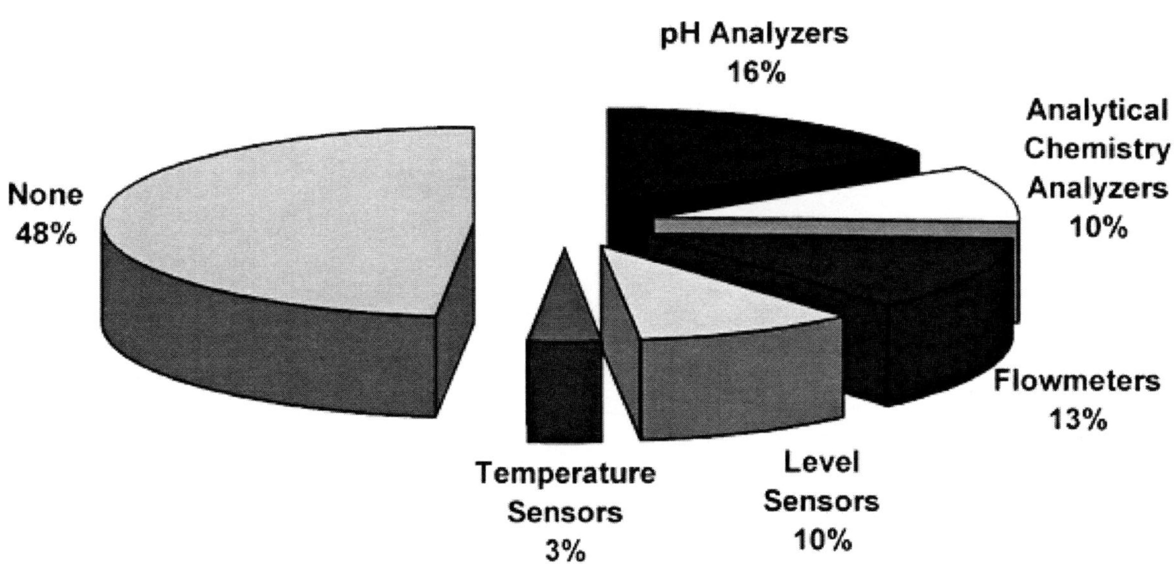

Figure 4-4. Instrumentation.

Automation and control redundancy is depicted in Figure 4-5. Redundant power sources (17%), generators (16%), and DCS/SCADA systems (16%) are reported by surveyed facilities as the primary automation and control equipment with redundancy. Alternative redundant design practices are shown by the use of multiple control strategies (6%) and dual field mounted control boxes (3%).

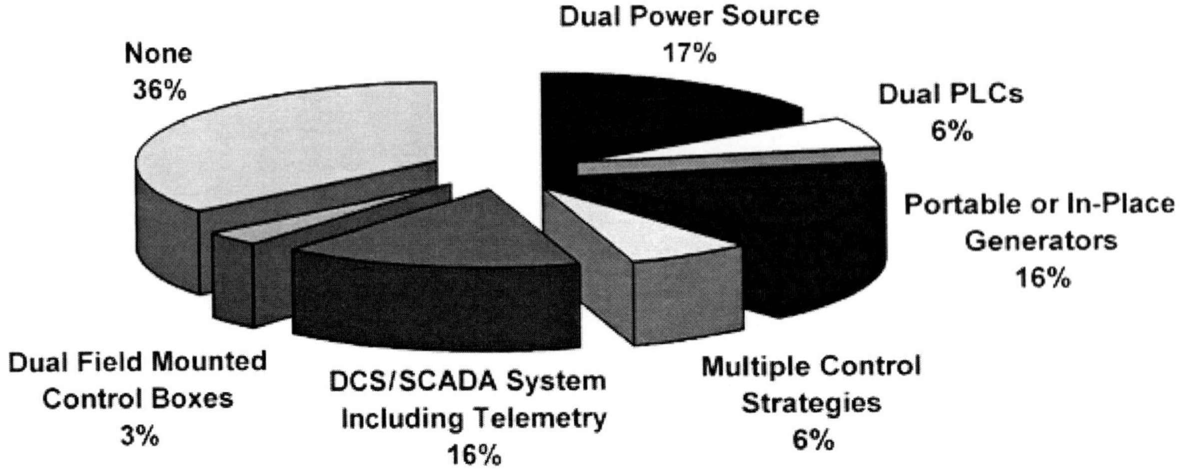

Figure 4-5. Automation and control.

4.4 Redundancy Reasons and Ratings

Treatment facilities with industrial pretreatment processes in place reported the primary reason and overall ratings of redundancy design practices for their plant as shown in Figures 4-6 and 4-7, respectively.

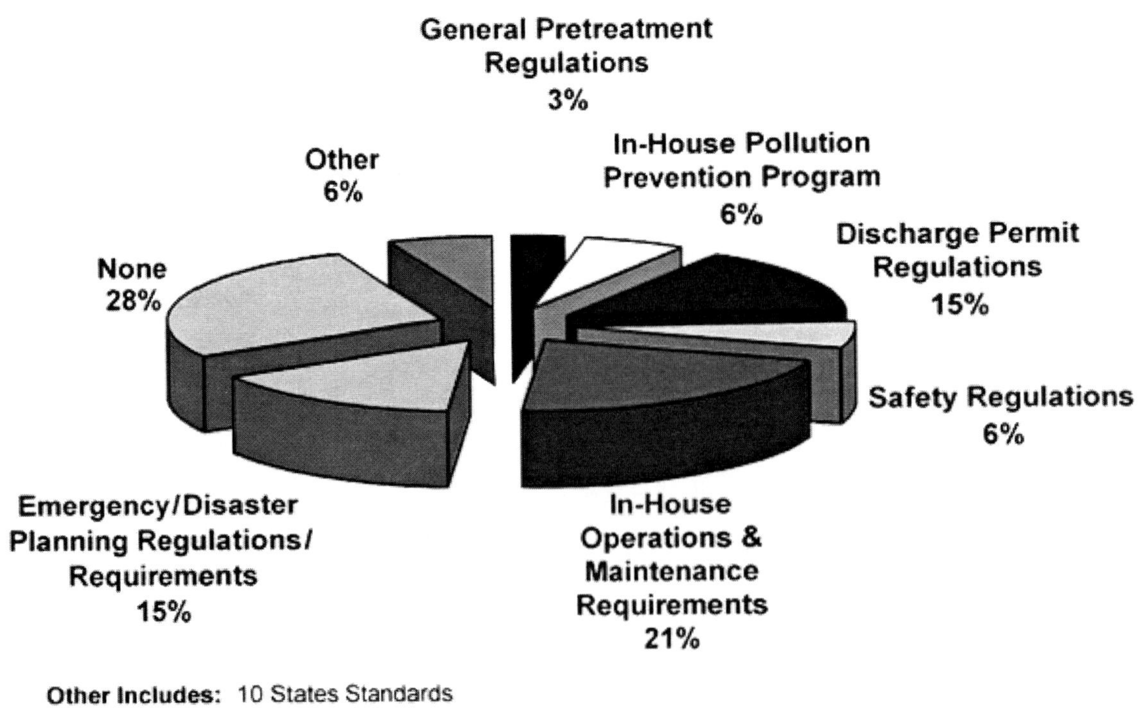

Figure 4-6. Reasons for redundancy.

In Figure 4-6, 21% of respondents reported that in-house operations and maintenance requirements are the primary driving force behind redundancy design practices. Other reasons for redundancy include discharge permits and emergency planning regulations, both reported by 15% of facilities.

As indicated by Figure 4-7, 42% of the reporting facilities rated their redundancy practices as extremely effective. A rating of somewhat effective (as defined as redundancy used for maintenance only in the project survey) was reported at 25%. In addition, 25% also stated that their redundant industrial pretreatment equipment was not currently in use, but may be used with future growth requirements. 8% of facilities indicated that their redundancy systems have never been used and they do not foresee a future use for these systems.

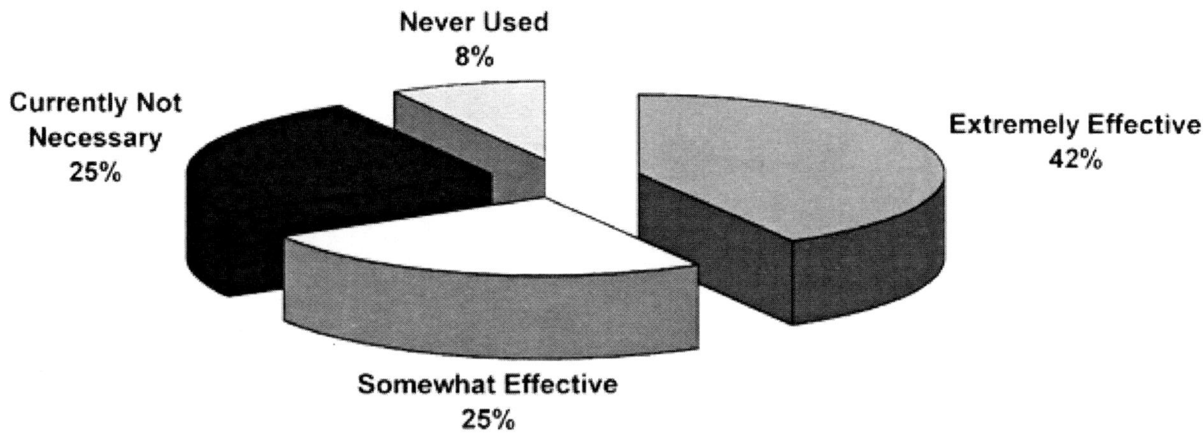

Figure 4-7. Redundancy ratings.

CHAPTER 5.0

PRELIMINARY TREATMENT

5.1 Introduction and Background

The preliminary treatment process, according to WEF MOP-8, handles the removal of troublesome components of sewage, such as large solids, rags, grit, odors, and peaks in hydraulic loading. By reducing the amount of these components in the wastewater stream, the preliminary treatment process adds to the effectiveness of downstream processes. However, the equipment used in the preliminary process is also subjected to the extremes of these troublesome components, making redundant equipment a necessity for providing a reliable preliminary treatment process. Appendix G provides detailed survey results in a tabular format for preliminary treatment, and results are summarized below.

5.2 Infrastructure and Equipment

The critical infrastructure for the preliminary treatment represents screening, grit removal, septage receiving, and flow equalization. Bypass units for grit removal account for 31% of the reporting facilities (Figure 5-1), and 28% reported having screening bypass units in place. Flow equalization bypasses were reported by 5%, and septage receiving bypasses were reported by 3%.

33% of the plants reported having no bypass capability, and 46% indicated that they had no infrastructure to provide backup capacity. Backup screening and grit removal infrastructure were reported by 21% of the surveyed facilities (Figure 5-2). 13% reported infrastructure that provides additional storage for wastewater received. Flow equalization backup and other redundant units were both reported by 10% of the facilities.

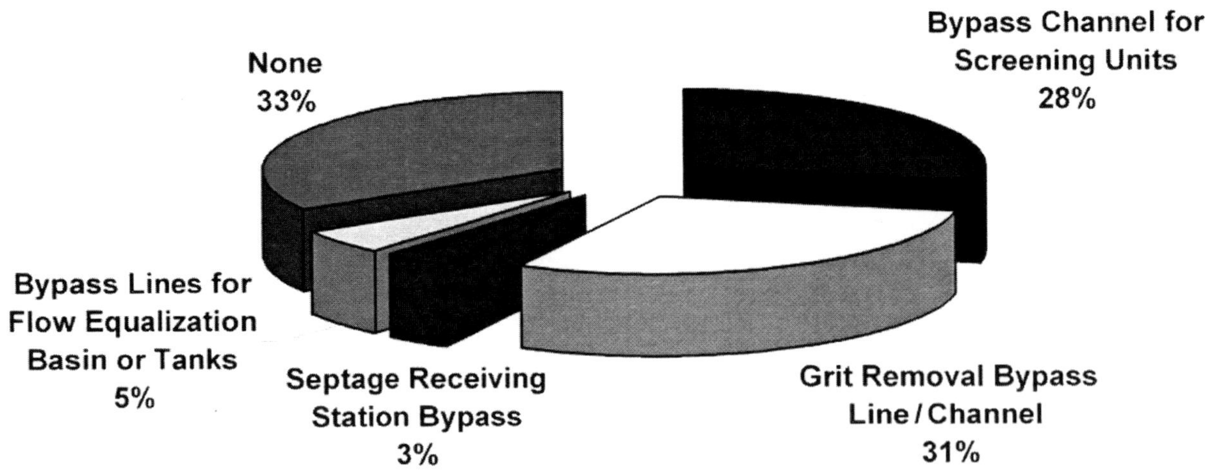

Figure 5-1. Infrastructure bypass units.

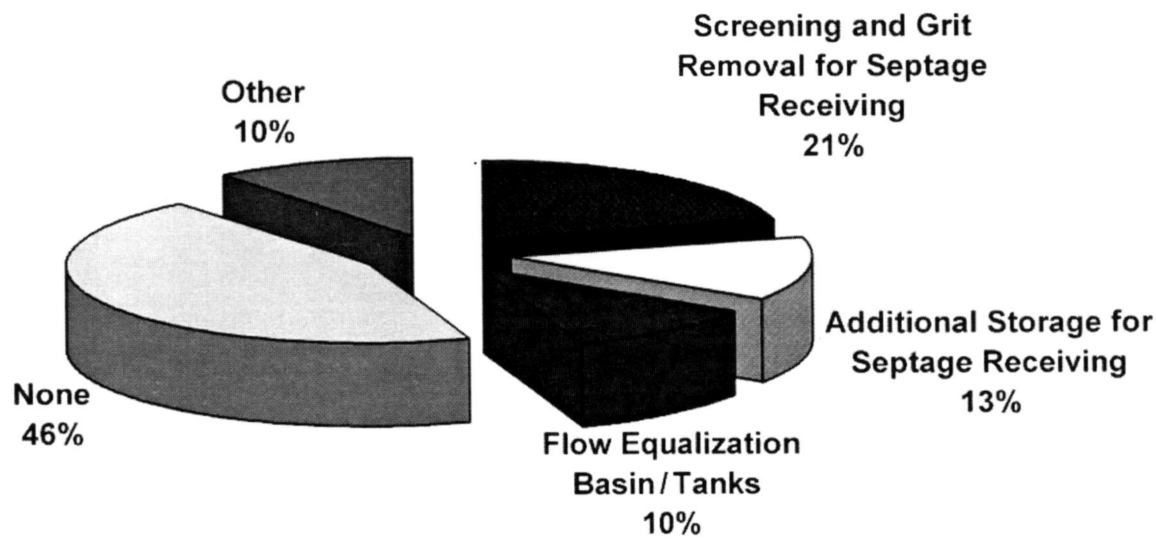

Other Includes: Take one train out of service in low flows; redundant grit tanks; redundant units; and multiple units.

Figure 5-2. Infrastructure backup units.

Treatment facilities reported bar screens (32%) and grit chambers (25%) with the highest redundancy for equipment in the preliminary treatment process. Bar screens experience extreme wear as the removal of large solids are processed, and grit chambers often fill and need frequent maintenance. Only 9% of the surveyed plants reported grinders and comminutors as redundant. Figure 5-3 shows that redundant equipment used in the preliminary treatment process varies greatly and is most likely dependent upon influent flow characteristics.

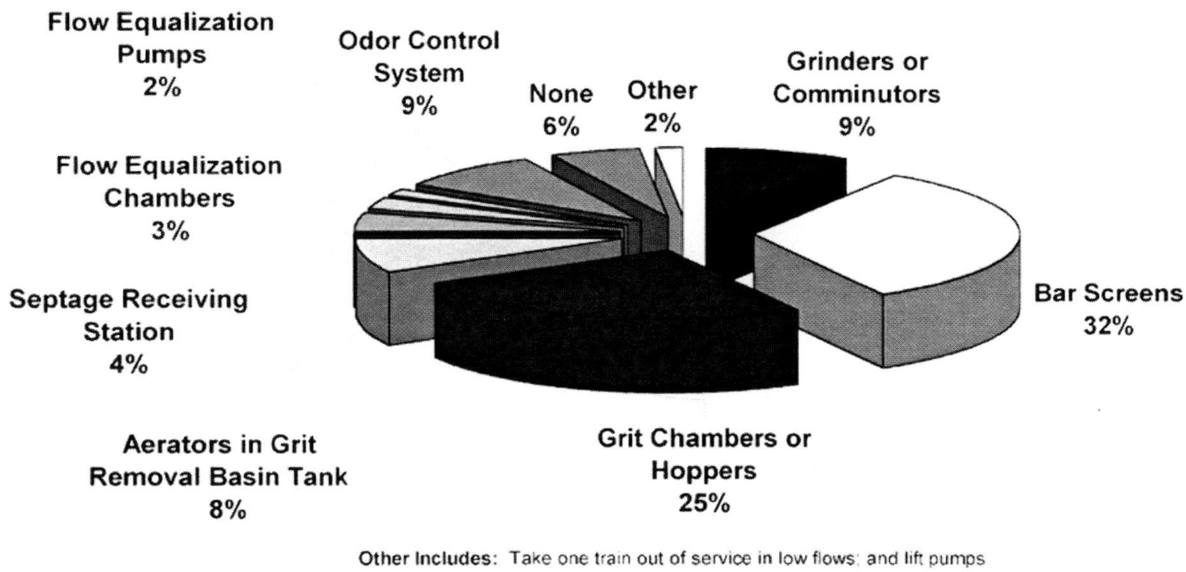

Figure 5-3. Equipment.

5.3 Instrumentation and Automation

As indicated in Figure 5-4, instrumentation redundancy in the preliminary treatment process seems to be split between those treatment plants that reported no redundant instrumentation (46%) and the remaining treatment plants that reported a spectrum of redundant instruments. The main instrument used redundantly for the preliminary treatment process is the level sensor in the bar screen channels, reported by 27% of plants. Few treatment facilities practice redundancy for their instrumentation for level sensors in the septage receiving basins (5%), level sensors in the flow equalization basins (4%), air flowmeters in the grit chambers (7%), pH analyzers in the septage receiving stations (5%), and flowmeters in the flow equalization basins (2%). Other redundant preliminary treatment process instrumentation (4%) included explosive gas meters, portable pH meters, dissolved oxygen (DO) meters, temperature instruments, and timer controls to backup the bar screens.

For preliminary treatment automation and control redundancy practices, Figure 5-5 displays that there are four main components reported by treatment facilities: dual power sources (29%), DCS/SCADA systems (23%), portable or in-place generators (16%), and multiple control strategies (16%). Figure 5-5 also shows that dual PLCs (6%) and dual field mounted control boxes (3%) are redundant in some facilities whereas approximately 3% of facilities reported having no redundancy for automation and control of the preliminary treatment process. Other automation and control redundant components for the preliminary treatment process (4%) include dual network loops, uninterruptible power supplies, and dual power sources.

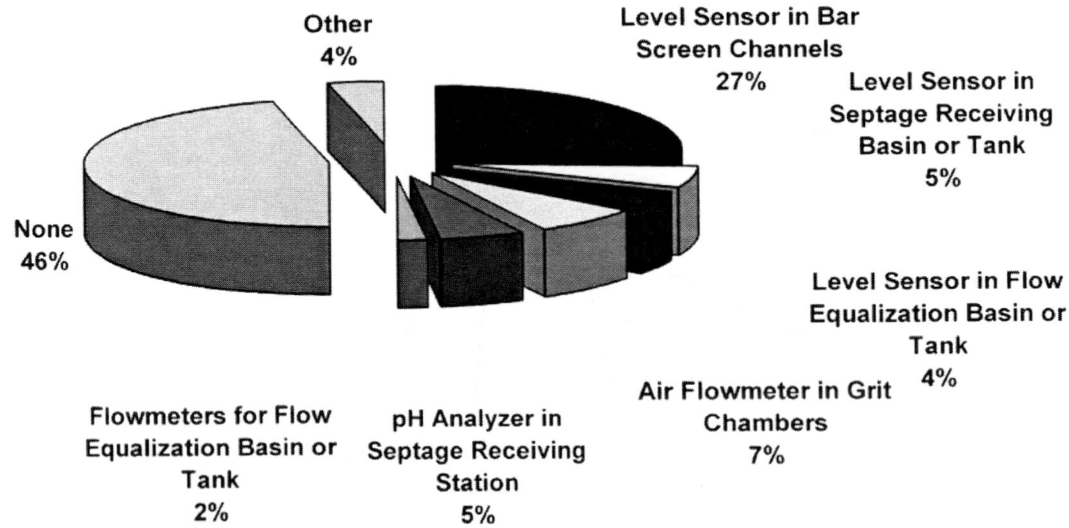

Other Includes: Explosive gas meter, portable pH, D.O. meter, temp and timer control to back up bar screen; and uninterruptible power supplies (AC with battery backups).

Figure 5-4. Instrumentation.

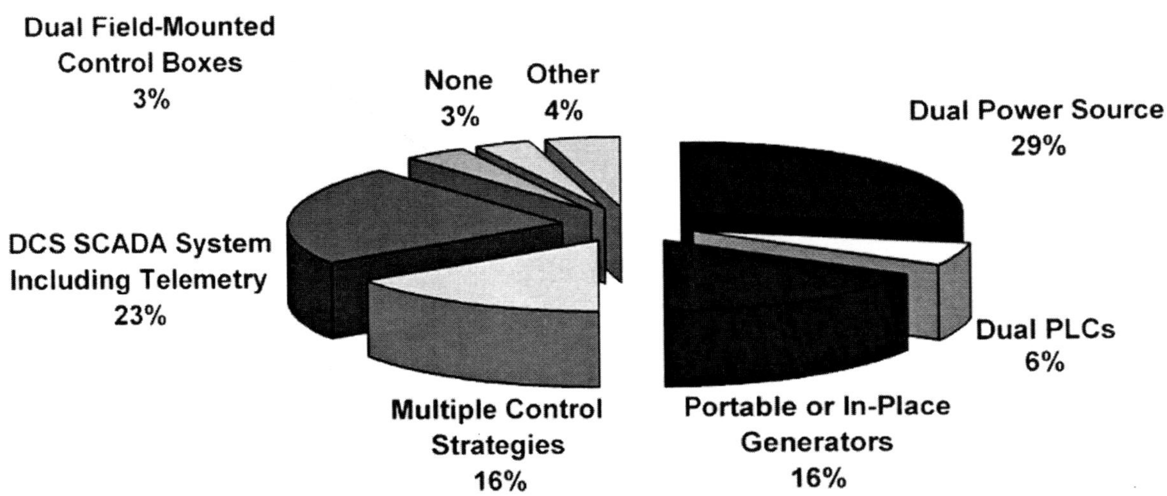

Other Includes: UPS; dual network loop; the treatment plant has a dual power source; 10-States Standard; and uninterruptible power supplies (AC with battery backups).

Figure 5-5. Automation and control.

5.4 Redundancy Reasons and Ratings

Figure 5-6 shows that the greatest reason for practicing redundant design in the preliminary treatment process is in-house operations and maintenance requirements (46%), followed by emergency/disaster planning regulations or requirements (23%). Safety accounted for 11% of the surveyed facilities, and 10% reported that state/provincial guidelines or regulatory requirements such as the NFPA 820, Water Resources Act, RCRA, and 10-States Standards were the reasons for redundancy design practices in the preliminary treatment process. Only 4% noted that the IEEE standard 493 was a reason for their redundancy, whereas 2% reported no redundancy.

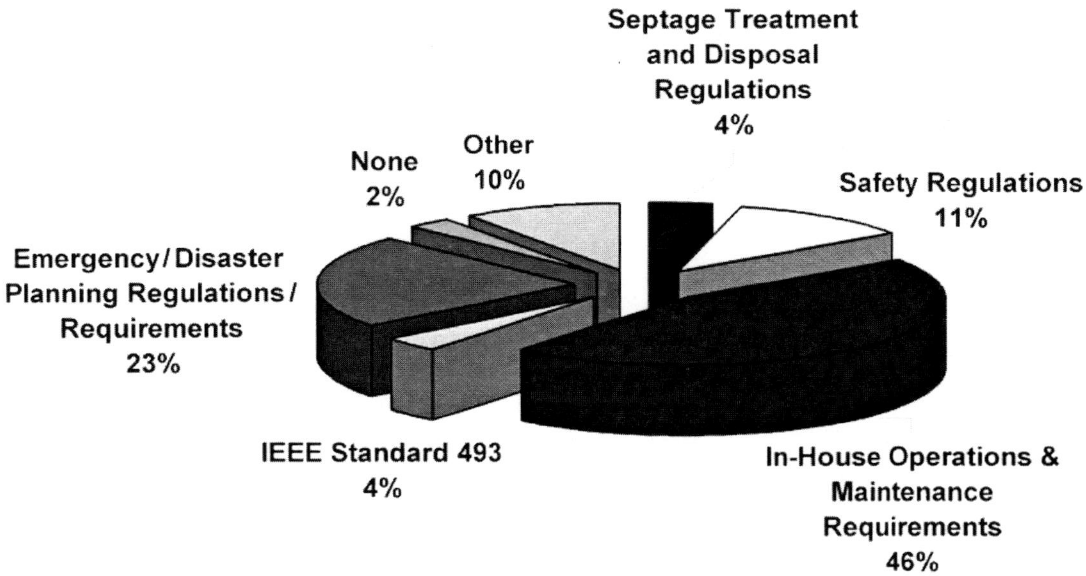

Other Includes: NFPA 820; Water Resources Act; RCRA; State Regulation; 10 States Standard; Federal, state and 10 state stds. for plant power redundancy; and State Standards.

Figure 5-6. Reasons for redundancy.

Figure 5-7 shows that surveyed treatment facilities reported that implementing redundancy design practices for preliminary treatment processes was extremely effective (53%) and somewhat effective (42%). Only 5% reported that redundancy design practices for the preliminary treatment process is not necessary. One facility noted that when it is necessary to take one unit from their preliminary treatment process out of service that they are capable of operating with less units for a short time, decreasing the need for redundancy, unless the units are down for a long period.

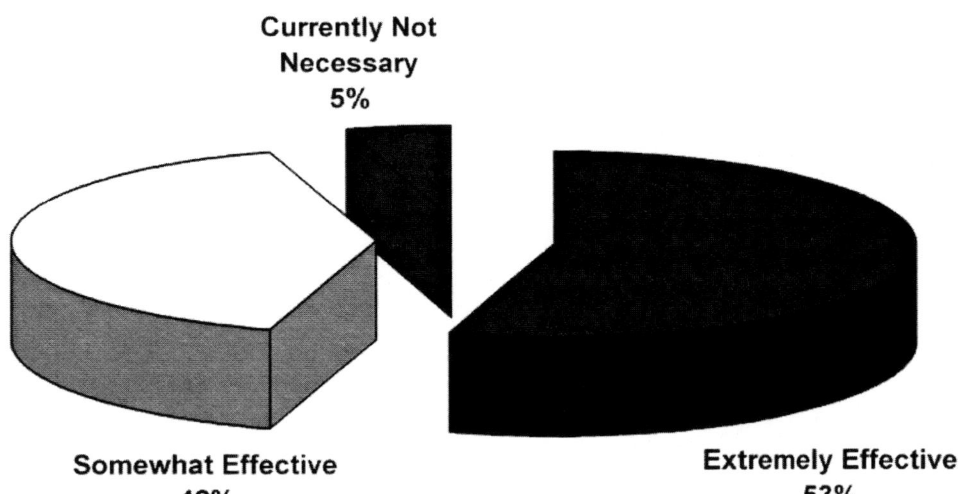

Comments: pH cross checks with multiple meters; When necessary to take one unit out of service we're capable of operating with lesser units for a short duration of time; No formal redundancy practices in-place; and None.

Figure 5-7. Redundancy ratings.

CHAPTER 6.0

Primary Treatment

6.1 Introduction and Background

The primary treatment process is one of the most widely used processes in wastewater treatment facilities in the United States, as noted in the EPA's 1988 Needs Survey. Sedimentation is the most frequently used method for primary treatment, while screening is occasionally used. The focus of the sedimentation process is to remove particulate and colloidal solids and to remove flocculent suspensions (Corbitt et al., 1989). Primary treatment generally uses clarification/sedimentation. However, for smaller systems, mechanical screens are used to substantially reduce suspended material in the influent before the primary treatment process begins.

The objective of primary treatment is to provide maximum suspended solids removal and thus, the concept of adding a chemical coagulant to the primary influent is an important operations option for maximizing suspended solids removal, particularly during high-flow periods. This primary treatment process can then produce an effluent that will allow further treatment by a biological process located downstream. Redundancy plays an important role in the efficiency of primary treatment that results in sufficient settled solids to be processed cost effectively by the wastewater facility. Appendix H provides detailed survey results in a tabular format for primary treatment.

6.2 Infrastructure and Equipment

The design of the primary treatment infrastructure (i.e., tanks, basins, and channels) depends upon the settling characteristics of the solids and the hydraulic capacity necessary for treatment at the maximum design flow expected. Infrastructure design, in turn, determines the primary treatment process equipment (pumps, collectors, skimmers, etc.) that will be required to support an efficient level of treatment.

Infrastructure redundancy reported by facilities (Figure 6-1) indicates that the majority of treatment plants (72%) do not have bypass units for their primary treatment process.

However, 28% of the surveyed treatment facilities have bypass systems for their sedimentation/settling tanks.

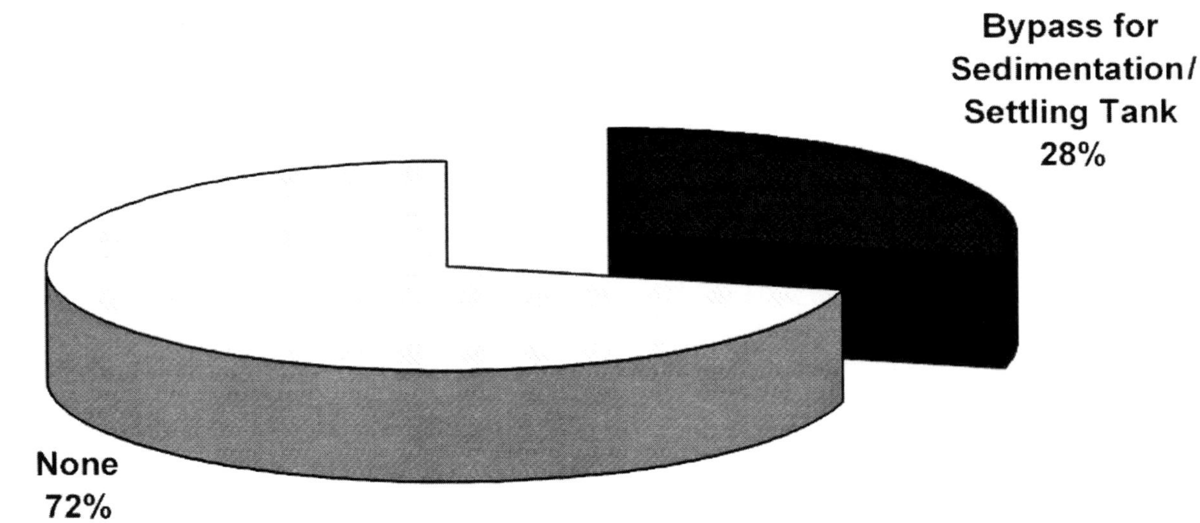

Comments: Bypass holding tanks only

Figure 6-1. Infrastructure bypass units.

Approximately half (55%) of the interviewed facilities noted having backup or secondary sedimentation/settling tanks for their primary treatment process (Figure 6-2). The remaining treatment plants (45%) reported not having backup/secondary or multipurpose units or noted that they took treatment trains out of service or had multiple units to deal with redundancy for maintenance events.

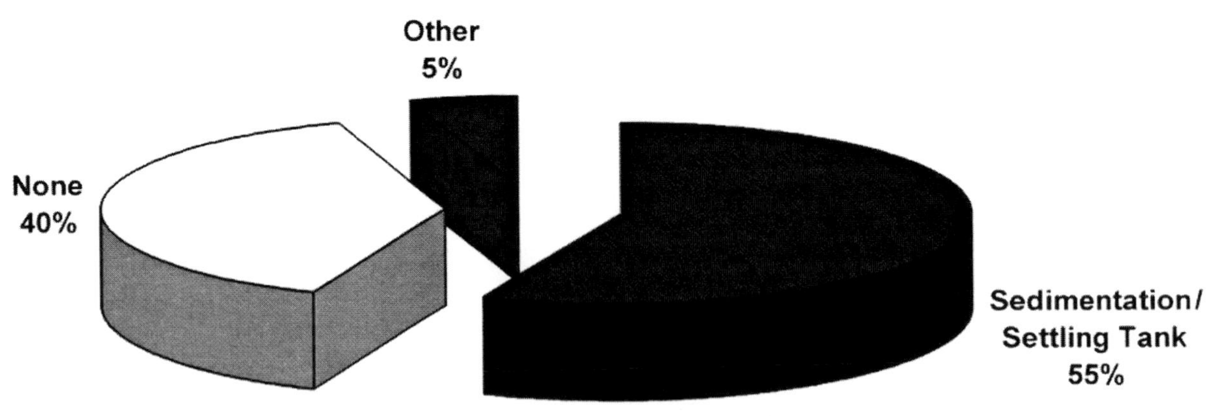

Other Includes: Take one train out of service in low flows. (winter/summer) and Multiple Units.

Figure 6-2. Infrastructure backup units.

Figure 6-3 shows that sludge pumps were reported as the most important piece of backup/secondary or multipurpose equipment (47%) followed by scum pumps (36%). Backup/secondary skimmers were reported by 9% of facilities, and only a few (3%) noted that they use backup chemical pumps or take their treatment train out-of-service in low flows. Five percent of the surveyed plants reporting having no backup/secondary or multi-purpose equipment for the primary treatment process.

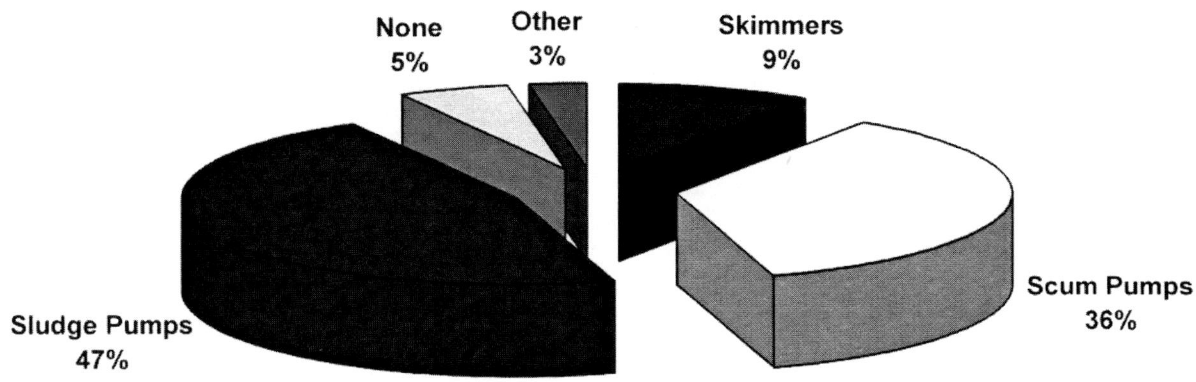

Other Includes: Take one train out of service in low flows. (winter/summer) and chemical pumps - polymer system.

Figure 6-3. Equipment.

6.3 Instrumentation and Automation

Figure 6-4 shows that more than half of treatment plants (58%) reported that they have no backup/secondary instrumentation for their primary treatment process. Only 20% reported using water-level sensors, and 10% reported having backup/secondary interface/sludge blanket level instruments. Backup/secondary suspended solids/sludge density meters and loss-of-flow sensors were both reported as redundant instrumentation by 6% of facilities.

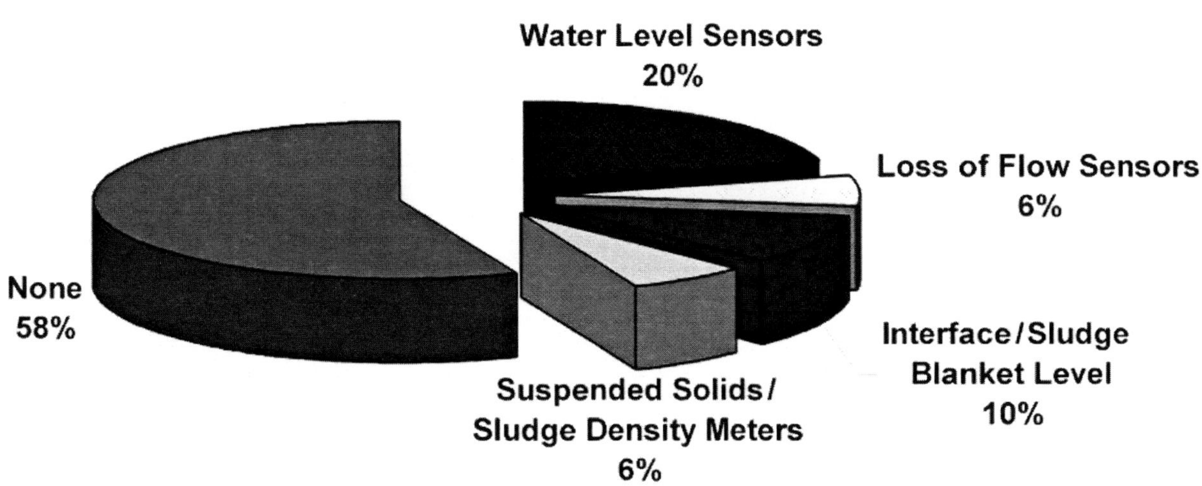

Figure 6-4. Instrumentation.

As shown in Figure 6-5, facilities reported the following redundant automation and control items for the primary treatment process: dual power source (26%) and DCS/SCADA systems (23%). In addition, portable or in-place generators are used as a backup or secondary or multipurpose unit by 17% of facilities, and multiple control strategies were reported by 15%.

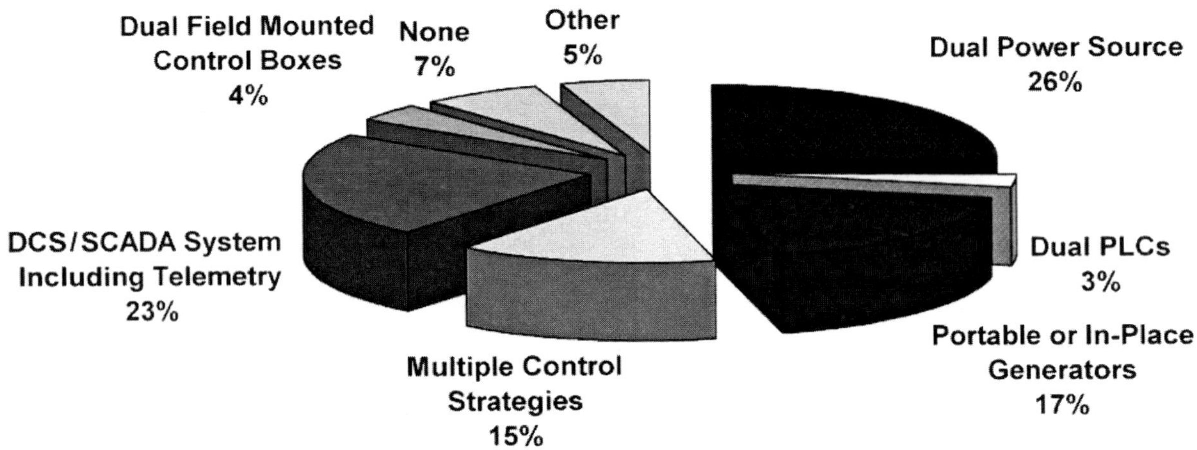

Other Includes: Polymer addition system; Dual network loop; 10 States Standard; and Uninterruptible power supplies (AC with Battery Backups).

Figure 6-5. Automation and control.

6.4 Redundancy Reasons and Ratings

The major reason for implementing redundancy in the primary treatment process was reported at 41% for in-house operations and maintenance requirements (Figure 6-6), followed by emergency disaster planning regulations/requirements (19%) and discharge permit regulations (19%).

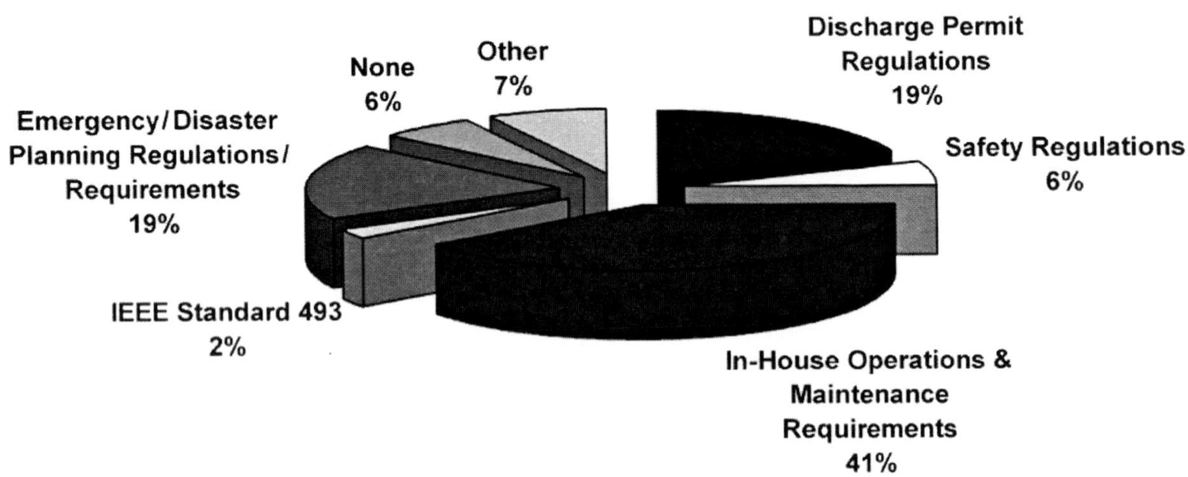

Other Includes: NFPA 820; Water Resources Act; 10 States Standard; Federal, state and 10 state stds. for plant power redundancy; and State Standards.

Figure 6-6. Reasons for redundancy.

Figure 6-7 provides the primary treatment redundancy ratings, which are reported as an almost even split between somewhat effective (48%) and extremely effective (47%). The remaining 5% of facilities noted that they never use redundancy.

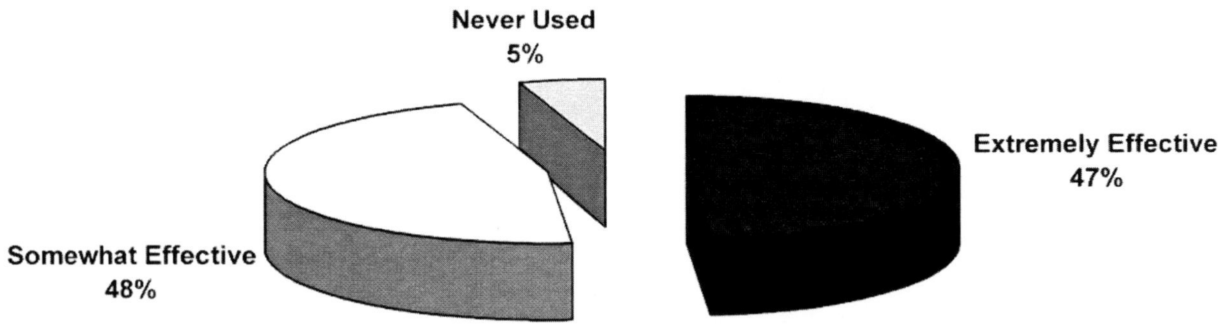

Figure 6-7. Redundancy ratings.

CHAPTER 7.0

SECONDARY TREATMENT

7.1 Introduction and Background

Secondary treatment, as described in the project survey form, characterizes the aerobic biological treatment process by either attached growth or suspended growth systems. This process can be accomplished by the use of several methods of treatment such as, rotating biological contactors (RBCs), trickling filters, activated sludge, and pond systems. The method selected for secondary treatment is often a requirement that is determined by the National Pollutant Discharge Elimination System (NPDES) permit for the wastewater treatment facility. The methods can also have variations in flow strategies and hydraulic retention times (HRT). Some basic flow strategies are completely mixed flow and others use plug flow. The flow strategies influence the HRT in order to optimize the treatment process. Appendix I provides detailed survey results in a tabular format for secondary treatment, and results are summarized below.

7.2 Infrastructure and Equipment

As the type of secondary treatment is determined, the level of redundancy becomes important to the process operations, infrastructure, and equipment used in treatment. For example, the activated sludge process may require multiple basins, mixed and/or plug flow strategies, air/oxygen compressors, mixing equipment, waste and return sludge pumping, and the ability for the process to have variable HRTs to accommodate variations in solids loading. The secondary clarifier systems, which follow the activated sludge process, have the requirement to settle out mixed liquor suspended solids (MLSS), produce a quality effluent, and accommodate hydraulic loading.

Figure 7-1 shows that 52% of the surveyed treatment plants reported having no secondary treatment bypass provisions, whereas 29% reported having secondary treatment bypass channels or piping, and 7% reported using oxygen bypass lines. An alternative to redundancy infrastructure bypass units was reported by 12% who noted ‘other,’ which included bypass return activated sludge line, taking one train out of service in low flows (winter/summer), influent bypass, effluent bypass, two oxygen plants interconnected, and a wet weather high flow bypass.

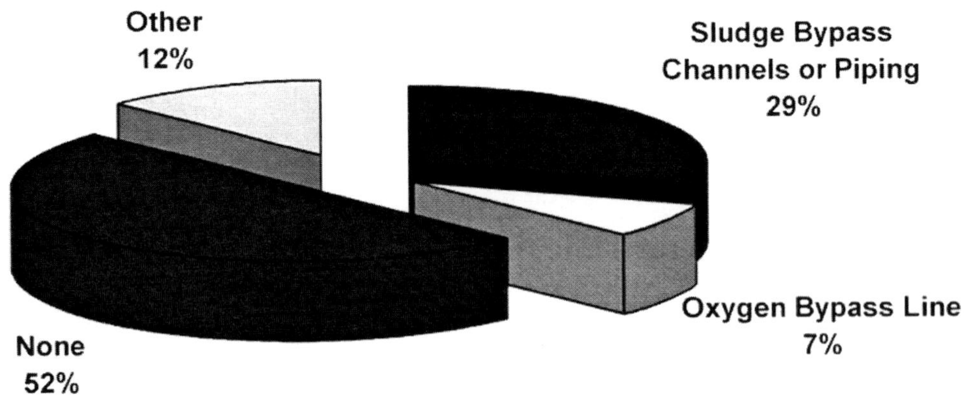

Other Includes: Bypass return activated sludge line; Take one train out of service in low flows (winter/summer); Influent bypass; 1 effluent bypass; 2 oxygen plants inter-connected; and wet weather high flow by-pass.

Figure 7-1. Infrastructure bypass units.

37% of facilities have backup/secondary suspended growth basins, 19% use backup secondary treatment process retention basins, and 13% use backup oxygen storage tanks (Figure 7-2). Only 4% reported having backup attached growth basins, whereas 20% reported no backup/secondary infrastructure for their secondary treatment process.

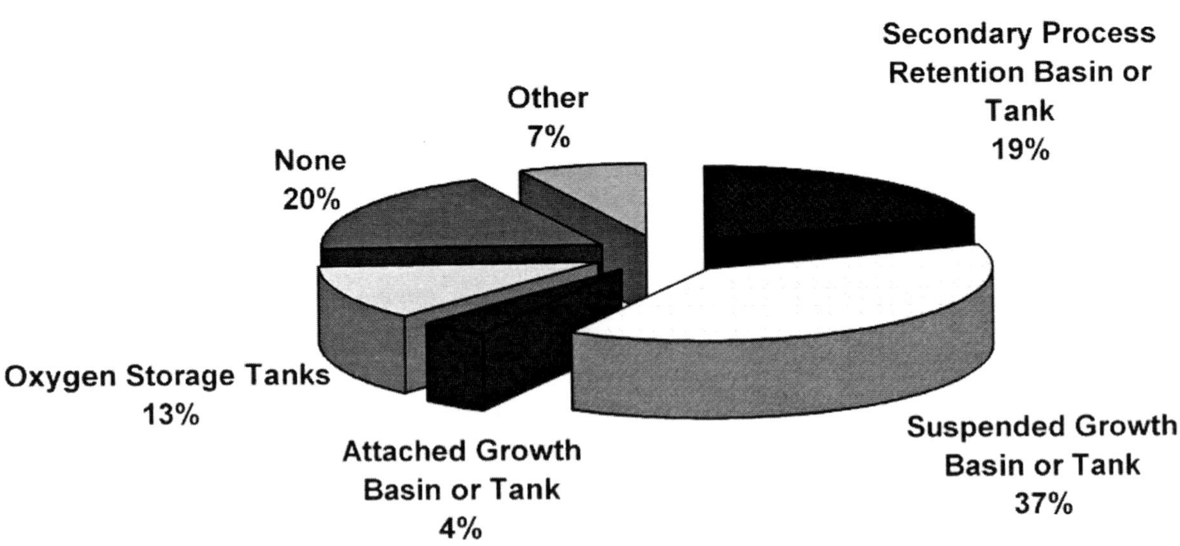

Other Includes: Take one train out of service in low flows. (winter/summer); clarifiers; and multiple units.

Figure 7-2. Infrastructure backup units.

For the secondary treatment process, pumps and blowers are reported as the most important redundant equipment (Figure 7-3): backup or secondary return activated sludge/waste activated sludge (RAS/WAS) pumps were reported by 26% of facilities, blowers were reported by 23%, and sludge pumping was reported by 21%. Chlorinators are used in a redundant configuration by 10% of facilities, and backup mixers and aerators are used by 8% and 6%, respectively. 4% reported having redundant air/oxygen compressors, and only 2% reported other backup equipment.

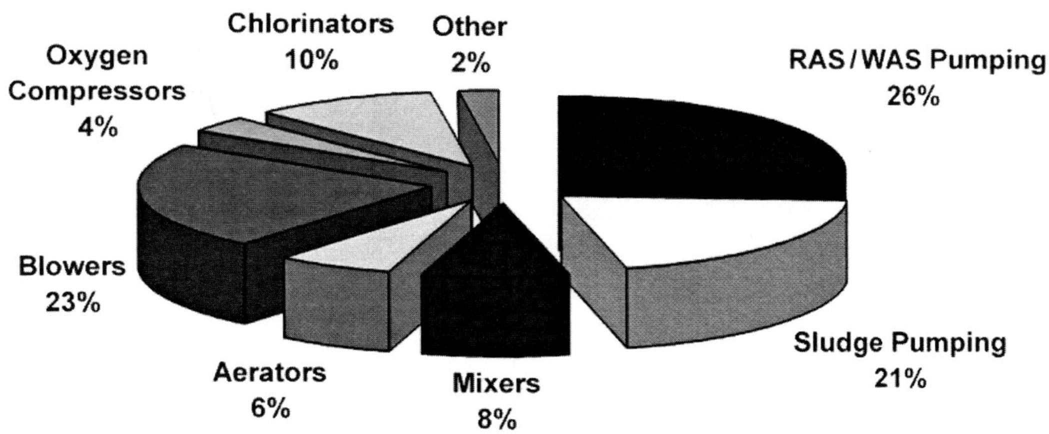

Other Includes: Take one train out of service in low flows. (winter/summer); process effluent pumps; multi-purpose influent feed pumps; and scum system.

Figure 7-3. Equipment.

7.3 Instrumentation and Automation

Figure 7-4 displays redundant instrumentation for the secondary treatment process. 28% of the plants reported that DO analyzers are the most critical instruments for their secondary treatment process. Facilities also reported using redundant RAS/WAS flowmeters (13%) and blower air flowmeters (12%). 10% reported using redundant interface/sludge blanket level instruments, and 8% noted they use backup suspended solids analyzers. 5% of facilities reported using redundant water-level sensors, and 4% reported using oxygen purity analyzers. Other redundant instrument technologies (2%) included biologic activity meters for nitrification and denitrification and airflow meters for a cryogenic plant. Alternatively, 18% of facilities reported that they do not have redundant instrumentation for their secondary treatment process.

Figure 7-5 displays that dual power sources (28%), backup DCS/SCADA systems (25%), and multiple control strategies (21%) are the primary redundant automation and control units used for the secondary treatment process. 12% of the plants have redundant portable or in-place generators for their secondary treatment process, and 5% have dual PLCs.

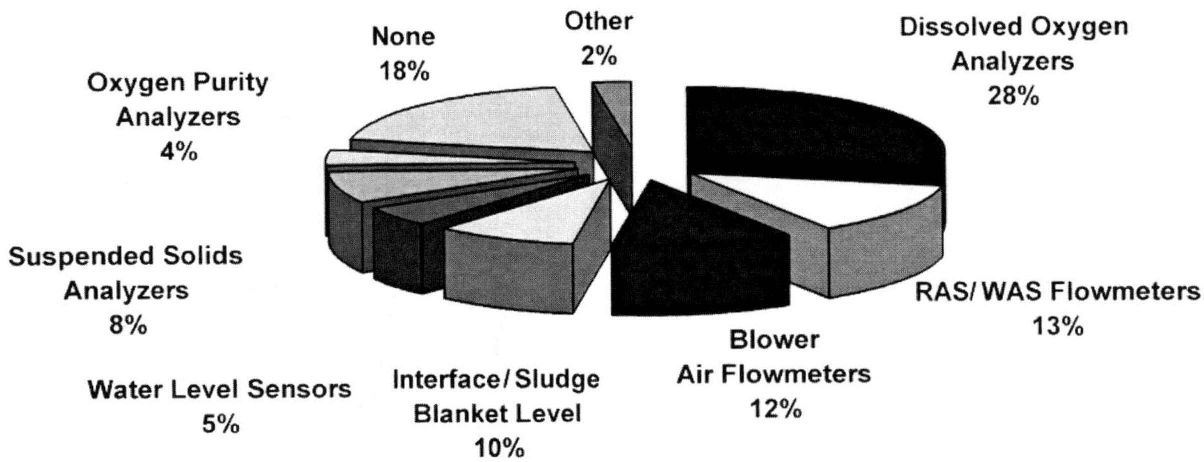

Other Includes: biologic activity meters for nitrification and denitrification; airflow meters are for the cryogenic plant; and may have spare units available on occasion.

Figure 7-4. Instrumentation.

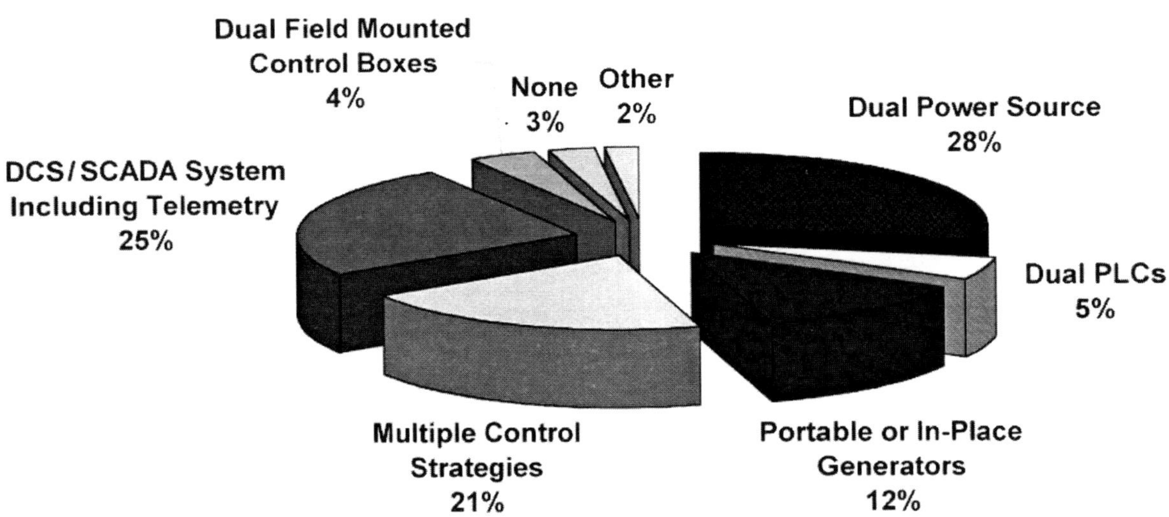

Other Includes: Uninterruptible power supplies (AC with battery backups)

Figure 7-5. Automation and control.

7.4 Redundancy Reasons and Ratings

Figure 7-6 shows that the most significant reason for redundancy design practices in the secondary treatment process is in-house operations and maintenance requirements (42%). 20% reported emergency/disaster planning, 18% reported discharge permit regulations, and 8% reported that safety regulations were the reasons for redundancy at their plant. Other reasons for redundancy (7%) included NFPA 820, Water Resources Act, 10-State Standards, good design practices, and state standards. Only 1% reported IEEE Standard 493 as the reason for redundancy, and 4% reported having no redundancy in their secondary treatment process.

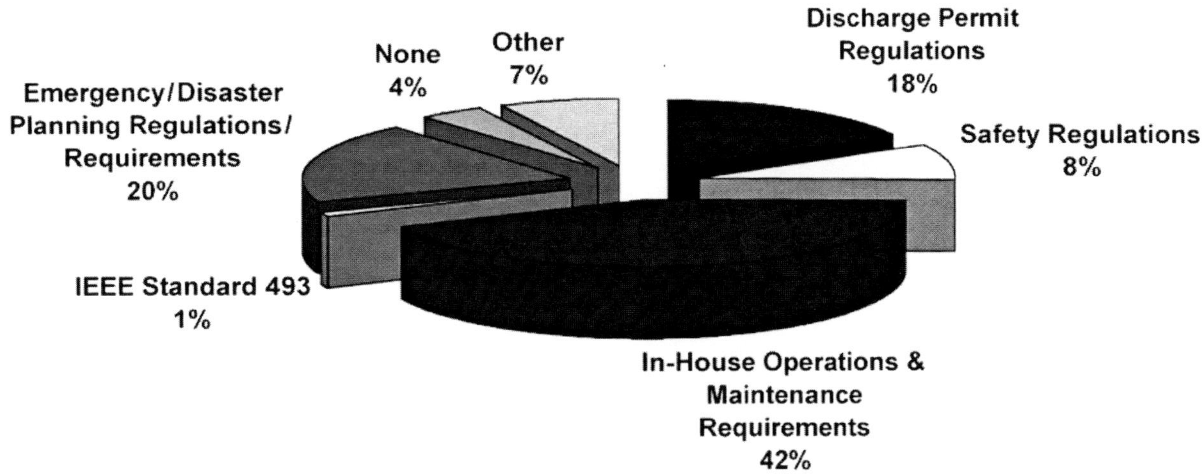

Other Includes: NFPA 820; Water Resources Act; 10 State Standards; good design practice; federal, state and 10 state standards for plant power redundancy; state standards.

Figure 7-6. Reasons for redundancy.

The primary reasons for secondary treatment redundancy noted in Figure 7-6 were rated as extremely effective (56%) and somewhat effective (42%) by most treatment plants (7-7). Only 2% reported that redundancy is not necessary for secondary treatment.

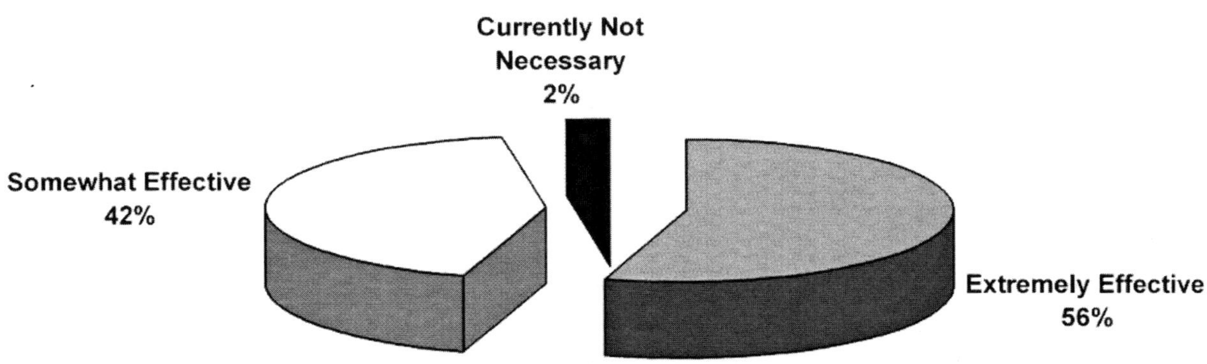

Comments: No formal redundancy practices in-place; Redundant storage tanks for LOX; and None.

Figure 7-7. Redundancy ratings.

CHAPTER 8.0

TERTIARY/ADVANCED TREATMENT

8.1 Introduction and Background

Tertiary and advanced treatment processes pose a significant challenge to wastewater treatment facilities. Tertiary treatment adds filters, tanks, chemical addition, and backwash equipment to the treatment process for final effluent water clarification or polishing.

Advanced treatment processes are becoming more essential to facilities that have to meet more stringent permit requirements that limit the amount of nitrogen and phosphorus discharged into surface and ground waters. Process designers rely on the use of microbial biochemistry and the activated sludge process environment to produce biological removal of nitrogen and phosphorus.

The advanced treatment process design can include tanks or basins that have aerobic, anoxic, and anaerobic treatment. In addition, the use of chemicals, waste or side stream pumps, and sludge handling equipment introduce a higher level of complexity to advanced treatment process operation and maintenance practices.

The treatment facility must carefully consider hydraulic capacities for various flows, retention times for the microbial biochemistry process to deliver adequate treatment, and incremental costs to the facility for treatment of sludge and recycle streams that are part of the tertiary and advanced treatment process. The reliability of these processes will depend on how redundancy is practiced. Appendix J provides detailed survey results in a tabular format for tertiary/advanced treatment, and results are summarized below.

8.2 Infrastructure and Equipment

Bypass filtration was reported as the highest rated bypass unit by 42% of the treatment plants for the tertiary/advanced treatment process (Figure 8-1). Bypass channels for secondary (that are used to divert any secondary treatment processes) were reported at 26%, whereas 29% reported having no redundant bypass units for their tertiary treatment process.

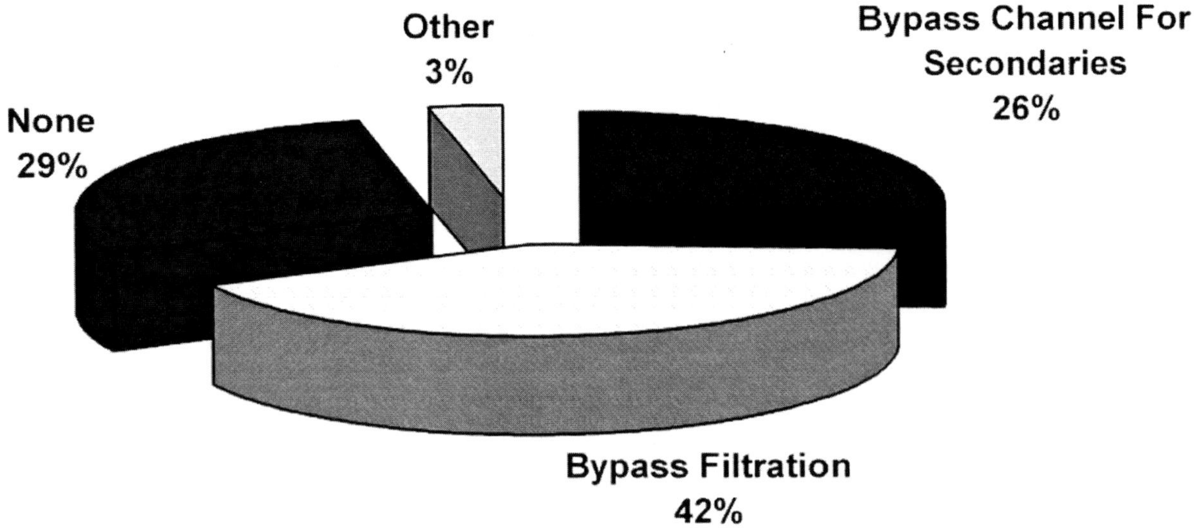

Other Includes: Not Applicable

Figure 8-1. Infrastructure bypass units.

Figure 8-2 displays the reported backup or secondary infrastructure units for the tertiary/advanced treatment process. 26% of the facilities reported having backup filters, 12% reported having redundant nutrient removal basins, and 9% reported having backup backwash tanks. Other (12%) reported redundant infrastructure units included wetlands cells and the alternative redundancy design practice of operating under design capacity permits bypass tanks on cells. 41% had no redundant tertiary/advanced treatment process infrastructure.

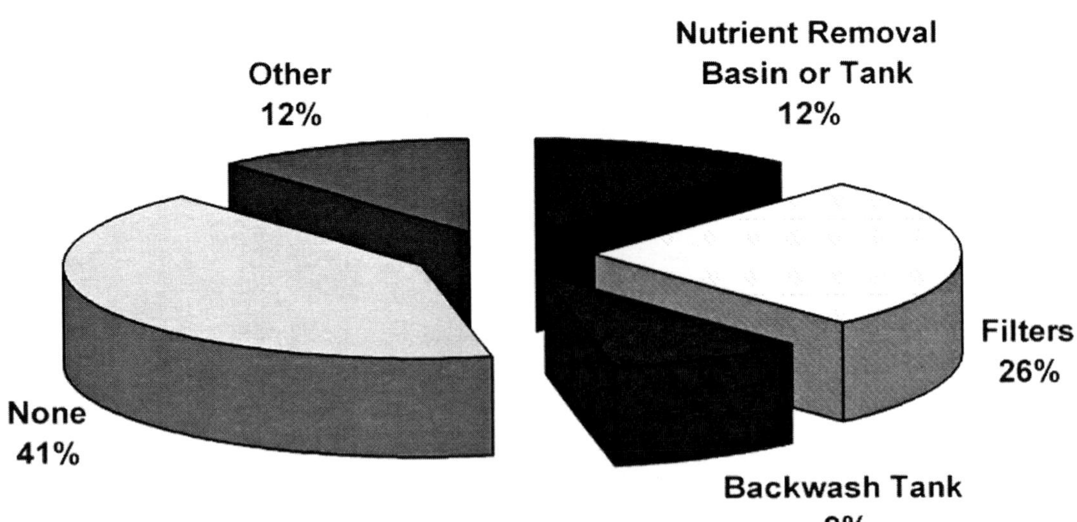

Other Includes: Operating under design capacity allows us to bypass tanks on cells; wetlands cell; not applicable; and multiple units.

Figure 8-2. Infrastructure backup units.

Facilities reported that pumps were the most important equipment to have in a redundant configuration for the tertiary/advanced treatment process. Figure 8-3 shows that backwash pumps accounted for 22%, sludge pumps 20%, and filter pumps 18%. In addition, 11% of facilities reported backup filter media as the most important equipment, 4% reported redundant odor control systems, and backup 2% each reported aerators and waste pumps.

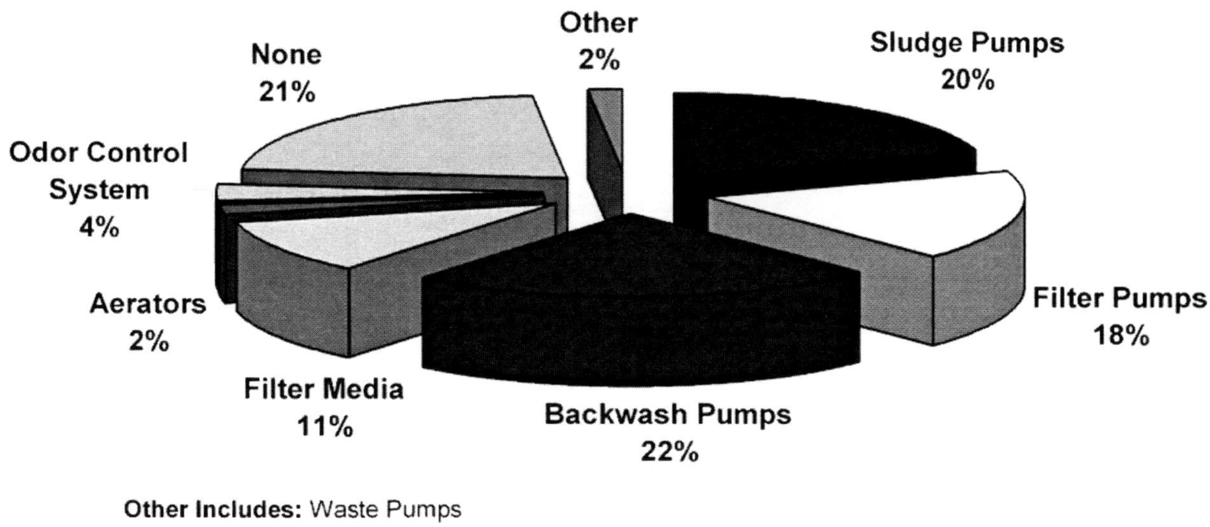

Figure 8-3. Equipment.

8.3 Instrumentation and Automation

Similar to the secondary treatment survey results, DO analyzers (19%) were the most frequently used redundant instruments for the tertiary/advanced treatment process (Figure 8-4). Level sensors were the second most frequently used redundant instrument (17%) followed by pH analyzers (12%) and flowmeters (7%). 2% of facilities each reported that ammonia, phosphorus, and nitrate/nitrite analyzers were the most frequently used redundant instruments. The "other" category was also reported by 2% of facilities and included turbidity meter, alkalinity testing, and portable analyzers. More than one-third of the facilities (37%) reported having no redundant instrumentation for their tertiary/advanced treatment process.

Figure 8-5 displays the reported backup/secondary automation and control units for the tertiary/advanced treatment process. Dual power source and DCS/SCADA systems were the most redundant automation and control units, used by 26% and 23% of treatment plants, respectively. Multiple control strategies (19%) and portable or in-place generators (17%) followed. 5% reported using dual field-mounted control boxes in a redundant fashion, 3% reported having dual PLCs, and 2% reported having backup uninterruptible power supplies. Only 5% reported having no redundancy for their tertiary/advanced treatment process automation and control.

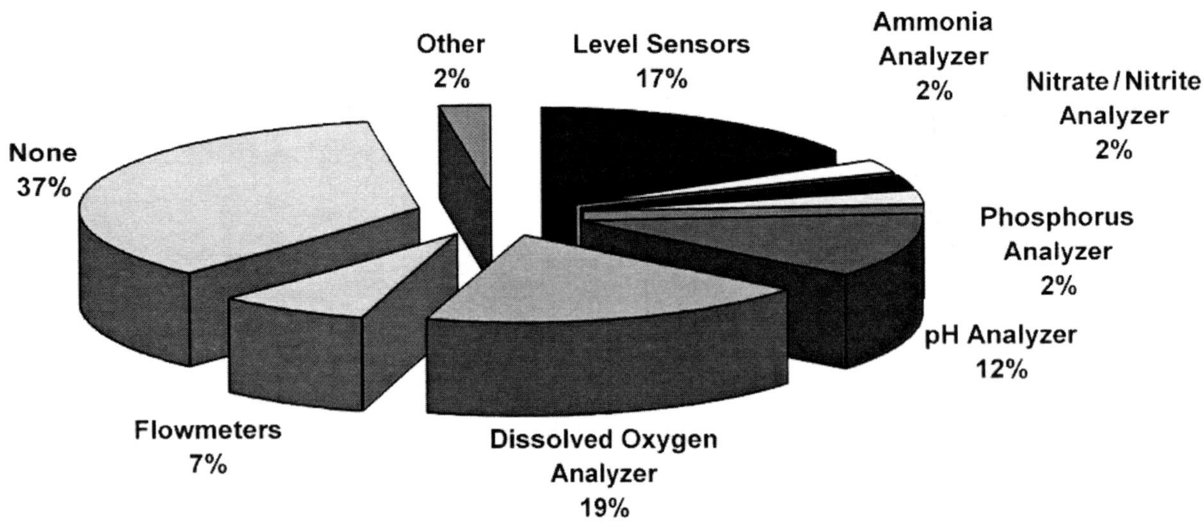

Other Includes: Turbidity meter; alkalinity testing; and portable analyzers.

Figure 8-4. Instrumentation.

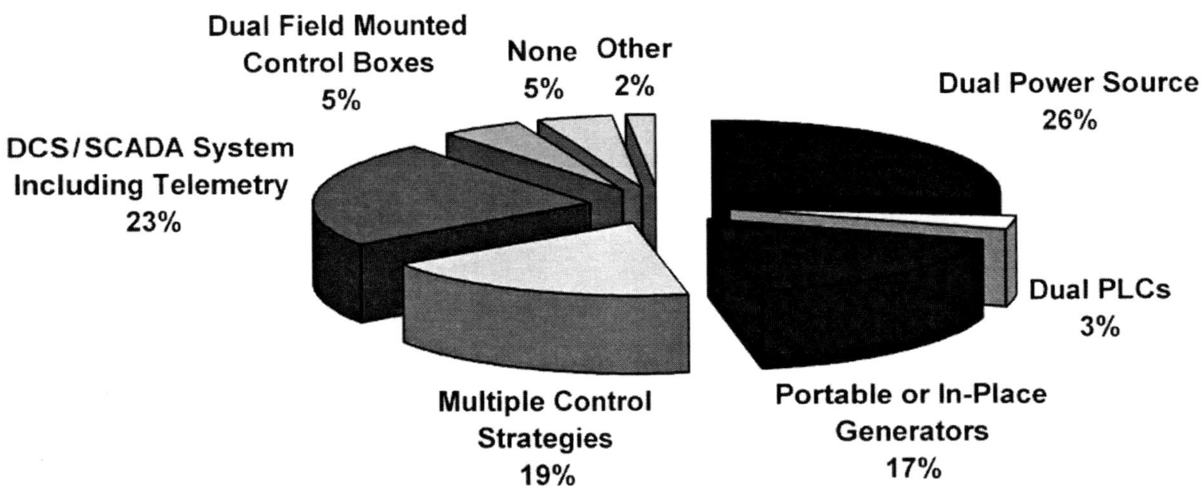

Other Includes: Uninterruptible Power Supply

Figure 8-5. Automation and control.

8.4 Redundancy Reasons and Ratings

Figure 8-6 displays the reported reasons for implementing redundancy for tertiary/advanced treatment. The most frequently reported reason was in-house operations and maintenance requirements (33%), followed discharge permit regulations (19%) and emergency/disaster planning regulations/requirements (19%). 8% of the plants recorded safety regulations as the main reasons for tertiary/advanced treatment redundancy, and 6% noted other reasons, including good design practices, federal, state, and 10-State Standards for power redundancy. Only 4% reported IEEE Standard 493 as one of the reasons for redundancy. 11% reported having no redundancy for their tertiary/advanced treatment process.

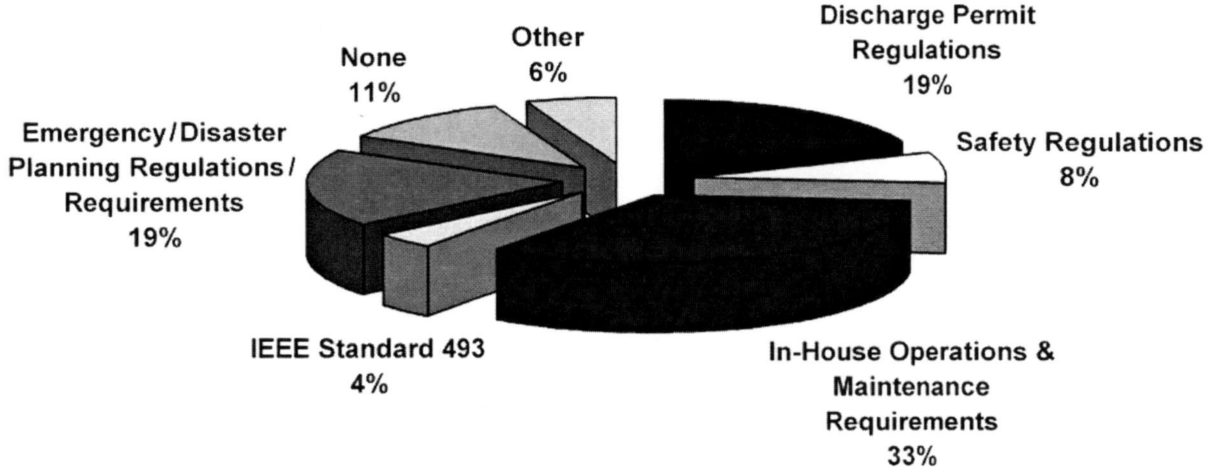

Other Includes: good design practices; federal, state and 10 state standards for plant power redundancy; and state standards.

Figure 8-6. Reasons for redundancy.

Figure 8-7 shows that the majority of the treatment facilities (88%) that practice redundancy design consider it effective. More than half (56%) reported redundancy as extremely effective, and 32% noted it is somewhat effective. Only 12% said that redundancy design practices for the tertiary/advanced treatment process currently are not necessary.

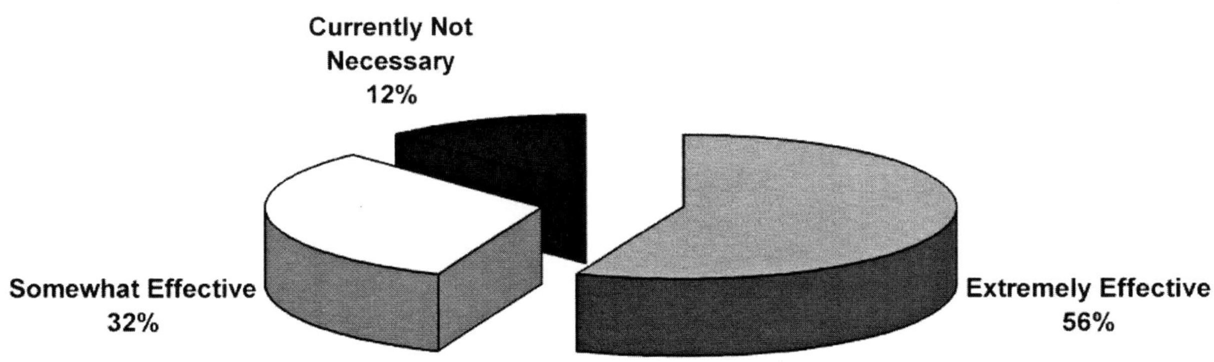

Figure 8-7. Redundancy ratings.

CHAPTER 9.0

Solids Processing

9.1 Introduction and Background

Solids processing treatment can vary at facilities depending upon the sources of sludge, (i.e., primary, secondary, mixed, and chemical). Corbitt et al. (1989) categorized the solids processing treatment into the following six processes: stabilization, conditioning, concentration, dewatering, drying, and residual disposal.

Within these six processes, the types of treatment will vary depending upon industrial and municipal waste flows, return waste, peak loads, and climatic conditions. For example, the stabilization process according to Corbitt et al. (1989) can include: aerobic or anaerobic digestion, chemical stabilization, composting, heat treatment, and use of lagoons.

Appendix K provides detailed survey results in a tabular format for solids processing, and results are summarized below.

9.2 Infrastructure and Equipment

The infrastructure attributed to the solids handling process has many components that depend on each for success. For example, a facility may have the anaerobic digestion, stabilization, chemical conditioning, centrifuges for dewatering, dissolved air flotation concentration, heat treatment drying, and land application for residual disposal. In addition, these components of the process require multiple pieces of equipment to provide for redundancy.

Facilities reported bypass lines for solids processing (34%) and digester gas bypass lines (33%) as the most frequently used redundant equipment for solids processing treatment (Figure 9-1). 6% of facilities reported other redundant bypass units for solids processing, including flare stacks, storage, and digested biosolids storage lagoons. 27% of facilities reported no redundant infrastructure bypass units for the solids processing.

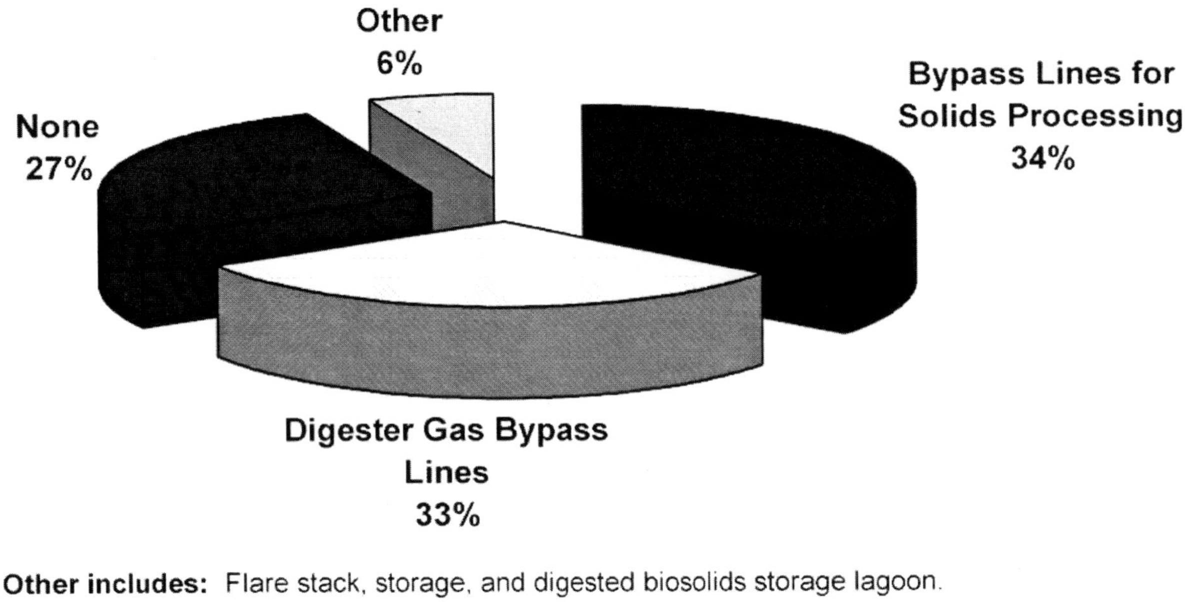

Figure 9-1. Infrastructure bypass units.

Figure 9-2 displays the numerous backup/secondary or multipurpose infrastructure units used in a redundant configuration for the solids treatment process. Digesters (17%) were the most frequently reported redundant infrastructure units for solids processing, followed by centrifuges, digester gas flares, and chemical feed tanks at 13% each. 11% reported having backup dissolved air flotation (DAF). 9% of facilities each reported having backup gravity belt thickeners and backup gravity belt presses. Only 5% reported having no redundant infrastructure for their treatment plants solids processing.

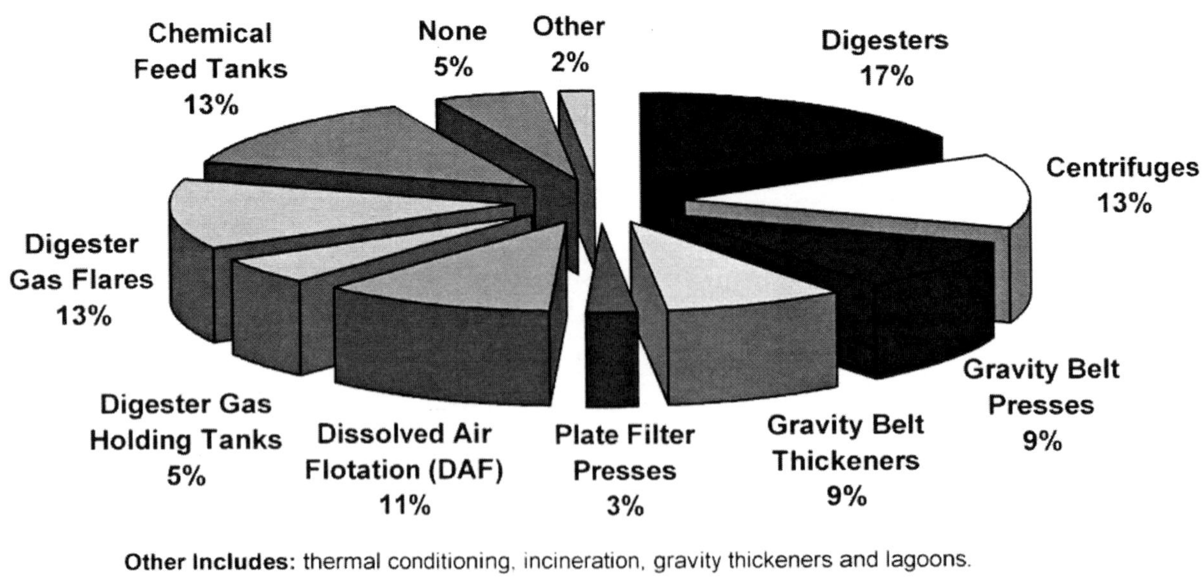

Figure 9-2. Infrastructure backup units.

Pumps are the most common redundant equipment used for solids processing at wastewater treatment facilities. Figure 9-3 shows that backup sludge pumps are recorded at 28% and backup chemical feeds pumps at 20%. 15% of plants use redundant boilers/steam supply, while 13% have both redundant gas compressors and chemical mixers. Odor control systems for solids processing are used in a redundant configuration by 7% of the surveyed facilities, and 1% reported having redundant heat exchangers. 3% of facilities reported having no redundant solids processing equipment at their treatment facilities.

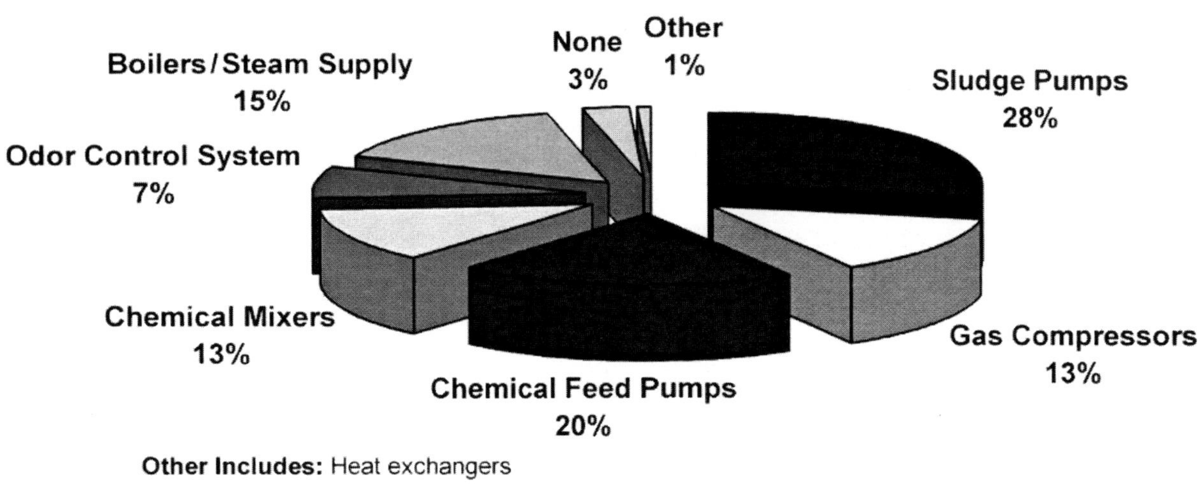

Figure 9-3. Equipment.

9.3 Instrumentation and Automation

Figure 9-4 indicates the variety of instrumentation used for wastewater solids treatment. Still, approximately 17% of treatment plants reported having no redundant instrumentation for their solids treatment processes. The most frequently used redundant instruments used for solids processing were level sensors (14%), followed by sludge flowmeters, chemical feed flowmeters and gas detectors at 12% each. 10% of facilities use backup temperature sensors, and 9% use backup gas flowmeters.

Figure 9-5 shows that dual power sources (26%) are the primary backup/secondary automation and control units used for solids processing, followed by DCS/SCADA (25%), and multiple control strategies (22%). Backup portable or in-place generators are used by 13% of treatment plants, and backup dual PLCs are used by 5%. Only 6% of treatment plants reported having no redundant automation and control units for their solids treatment process.

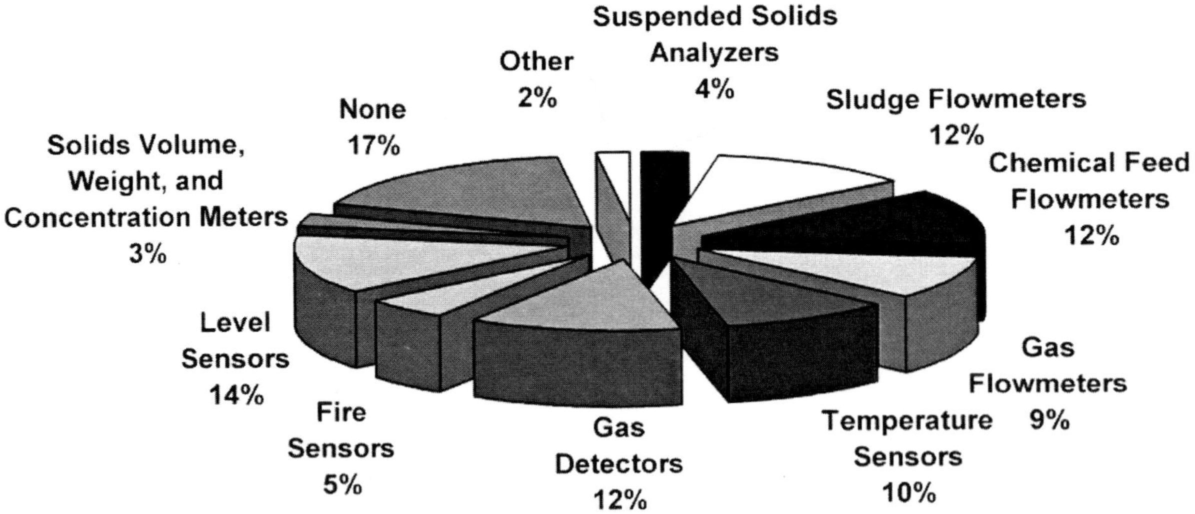

Other Includes: Multiple process trains can be strategically isolated; and nuclear density gauges for sludge.

Figure 9-4. Instrumentation.

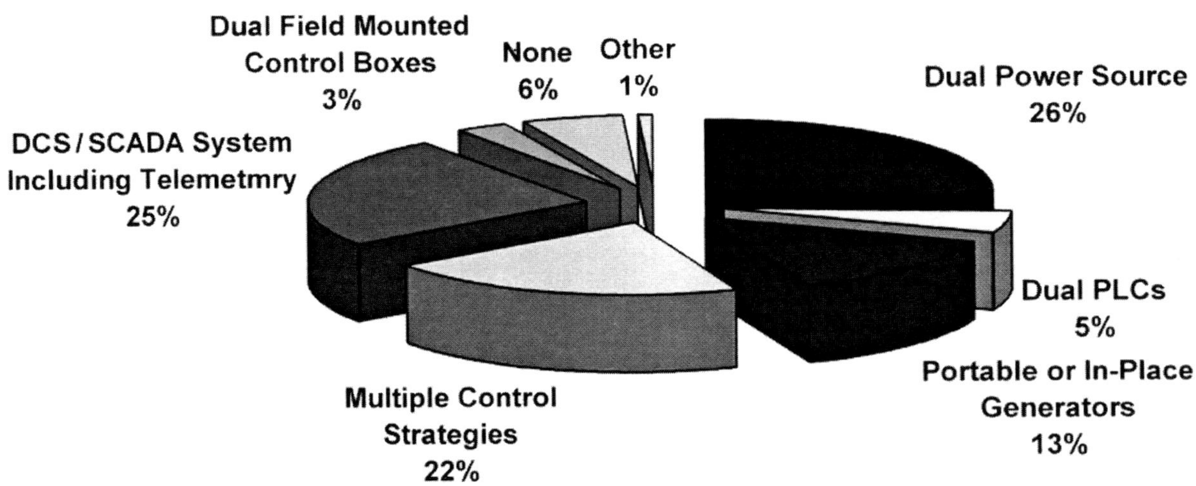

Other Includes: Uninterruptible power supplies (AC with battery backups)

Figure 9-5. Automation and control.

9.4 Redundancy Reasons and Ratings

In-house operations and maintenance requirements (37%) are the most common reason for implementing redundancy for wastewater solids processing (Figure 9-6), followed by solids processing regulations (16%), emergency/disaster planning regulations/requirements (16%) both at 16%, discharge permit regulations (13%), and safety regulations (9%). 5% of plants reported other reasons for redundancy, including NFPA 820, Water Resources Act, good design practices, and state standards. Only 1% reported the IEEE Standard 493 as one of the reasons for implementing redundancy for solids processing. 3% reported that they have no redundancy for solids processing.

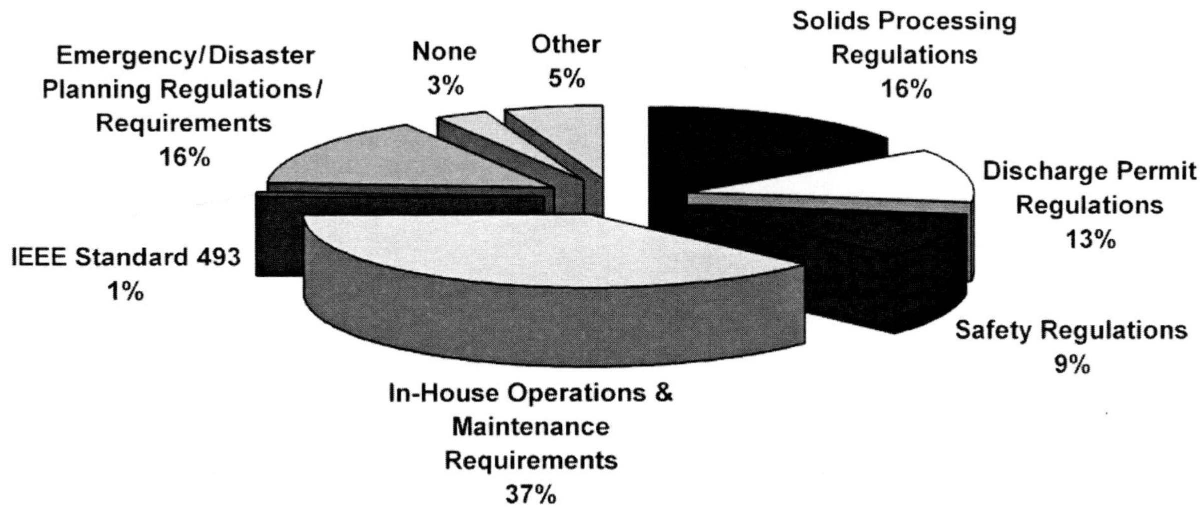

Other Includes: NFPA 820; Water Resources Act; Good Design Practices; and State Standards.

Figure 9-6. Reasons for redundancy.

Figure 9-7 displays the effectiveness ratings of implementing solids processing redundancy at wastewater treatment facilities. 57% rated solids processing redundancy as extremely effective, and 40% reported it as somewhat effective. Only 3% reported that redundancy design practices at their treatment facilities are not necessary.

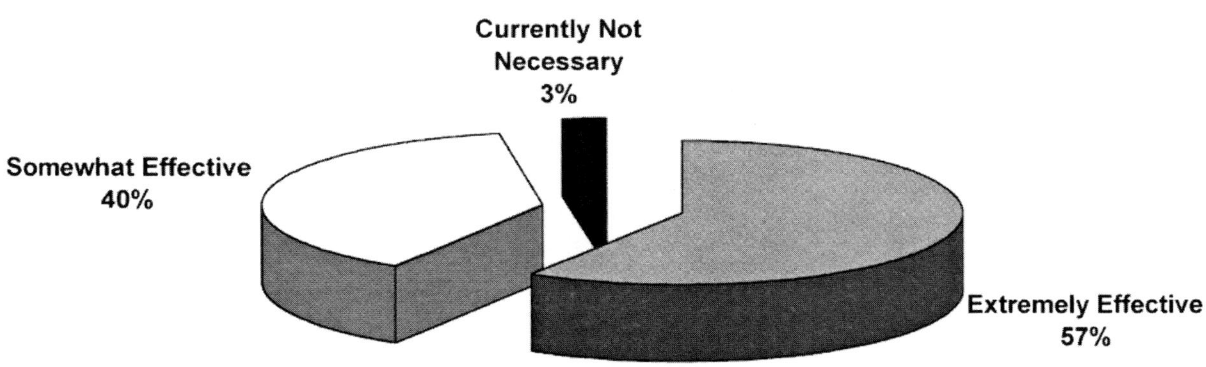

Figure 9-7. Redundancy ratings.

CHAPTER 10.0

DISINFECTION AND EFFLUENT

10.1 Introduction and Background

Disinfection processes are responsible for the destruction of disease-causing organisms in wastewater effluent (Corbitt et al., 1989). The Federal Water Pollution Control Act Amendments of 1972, per WEF MOP FD-10, unified the basics for determining disinfection treatment standards. One of the most common disinfection processes uses chlorine to destroy microorganisms. Ultraviolet (UV) light also is an effective means of disinfection.

The level of disinfection required by a facility is mandated by the appropriate state/provincial agency. The federal agencies provide guidelines for the receiving water to the states. The discharge permit and the possibility of fines for violators of disinfection guidelines influences the use of redundancy for the disinfection process. Appendix L provides detailed survey results in a tabular format for disinfection and effluent.

10.2 Infrastructure and Equipment

Chlorine used for disinfection must be distributed consistently throughout the effluent flow to maintain effectiveness. Contact basins, long effluent pipes, and mixing equipment are often used to insure adequate time for the chlorine to perform disinfection. Toxic byproducts emerge from chlorination, and many states also require dechlorination through the use of sulfur dioxide or sodium bisulfite. Chemical storage and equipment for the chlorination/dechlorination process requires redundancy not only for reliable process control but also to protect the safety of the operations and maintenance staff.

Ultraviolet light recently has become more prevalent in wastewater disinfection. Facilities need to consider the capacity of the UV system to handle variances in hydraulic flow as well as changes in turbidity, color, and organic compounds. Standby backup UV light sources and contact channels will be required by regulatory agencies (Corbitt et al, 1989).

Figure 10-1 indicates that the majority of treatment plants (43%) do not have bypass units for their disinfection and effluent treatment processes. However, 48% of treatment facilities have bypass systems for their effluent (24%) and/or disinfection channels or pipes (24%). In addition, 9% reported other alternatives to redundant bypass infrastructure units, including: using multi-pass channels that can change flow to dewater one-third of the tankage, temporarily installing disinfection in the bypass pipe, taking one train out of service in low flows, (i.e., winter/summer), and having multiple disinfection channels.

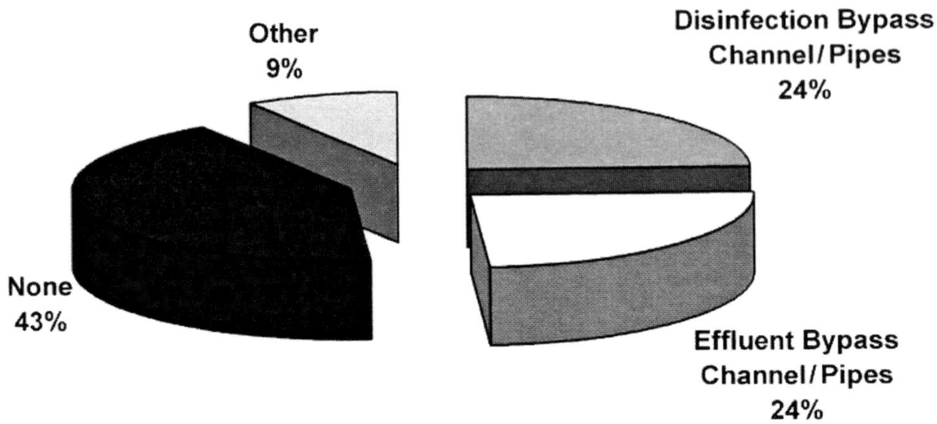

Other Includes: Multi-pass channel that can change flow to dewater one third of the tankage; Temporary installation of disinfection is installed in the bypass pipe; Take one train out of service in low flows. (winter/summer); and Multiple disinfection channels.

Figure 10-1. Infrastructure bypass units.

For backup/secondary infrastructure of disinfection and effluent, Figure 10-2 shows that the majority of treatment facilities (33%) reported using disinfection basins or tanks for backup or multipurpose units, followed by redundant final clarifier tanks (12%), backup aeration tankage (8%), redundant effluent holding tanks (6%), and backup liquid oxygen and vaporization storage tanks (4%). Other backup/secondary disinfection and effluent infrastructure (6%) includes multiple disinfection channels and acid neutralization tanks.

Figure 10-3 shows that chemical feed pumps (26%), chlorinators (22%), dechlorinators (16%), and effluent pumps (15%) are the primary redundant equipment used for the disinfection and effluent treatment processes. Chemical mixers are redundant at 8% of the surveyed treatment facilities, UV lamps at 4%, and backup or secondary air scrubbers at 3%. Other backup/secondary equipment was reported by 2% of the treatment plants and includes sulphonators, liquid chlorine injection valves, chemical storage tanks, and spray bar at dosage point. Only 4% of the treatment plants reported that they have no redundant equipment for their disinfection and effluent treatment processes.

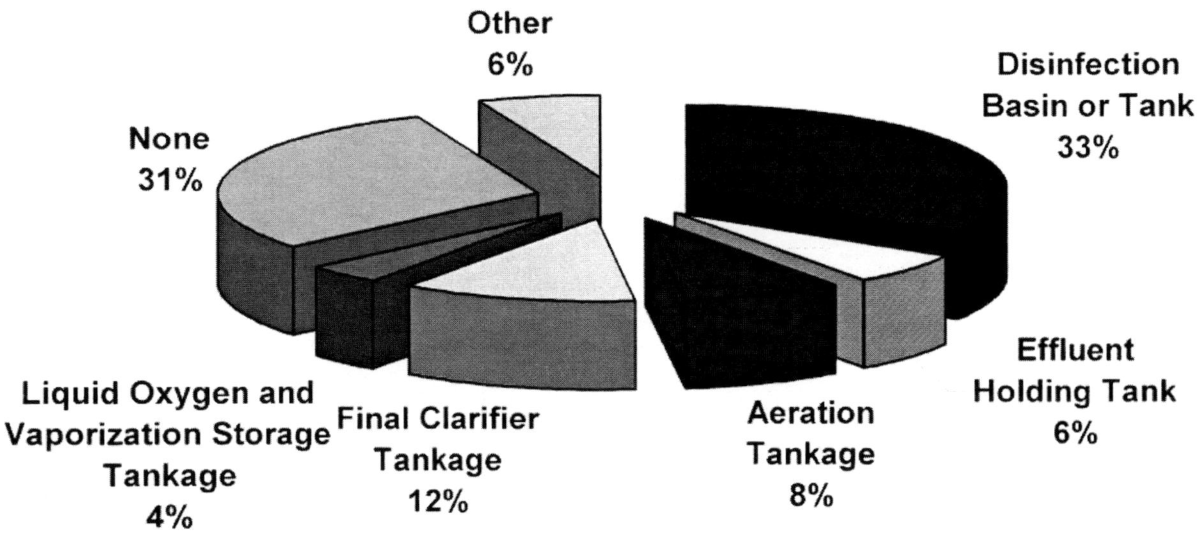

Other Includes: Multiple Disinfection Channels; Acid Neutralization Tank; and Multiple Units.

Figure 10-2. Infrastructure backup units.

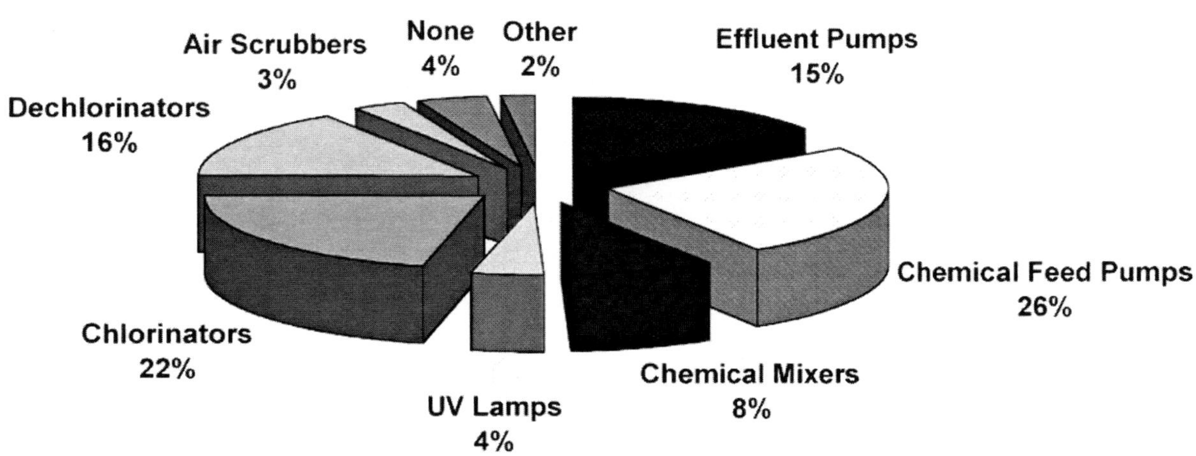

Other Includes: Sulphonators; liquid chlorine injection valves; and chemical storage tanks, spray bar at dosage point.

Figure 10-3. Equipment.

10.3 Instrumentation and Automation

Redundant instrumentation for disinfection and effluent treatment (Figure 10-4) includes chlorine analyzers (21%), followed by flowmeters (14%), and pH analyzers (13%). Dechlorination analyzers are redundant at 9% of facilities, and gas detectors, turbidity analyzers, and level sensors are each redundant at 8%. Backup or secondary temperature sensors, UV transmittance analyzers and other instruments, including portable analyzers and DO analyzers, are redundant at 1% of treatment plants. Approximately 16% of the facilities reported that they have no redundant instrumentation for their disinfection and effluent treatment.

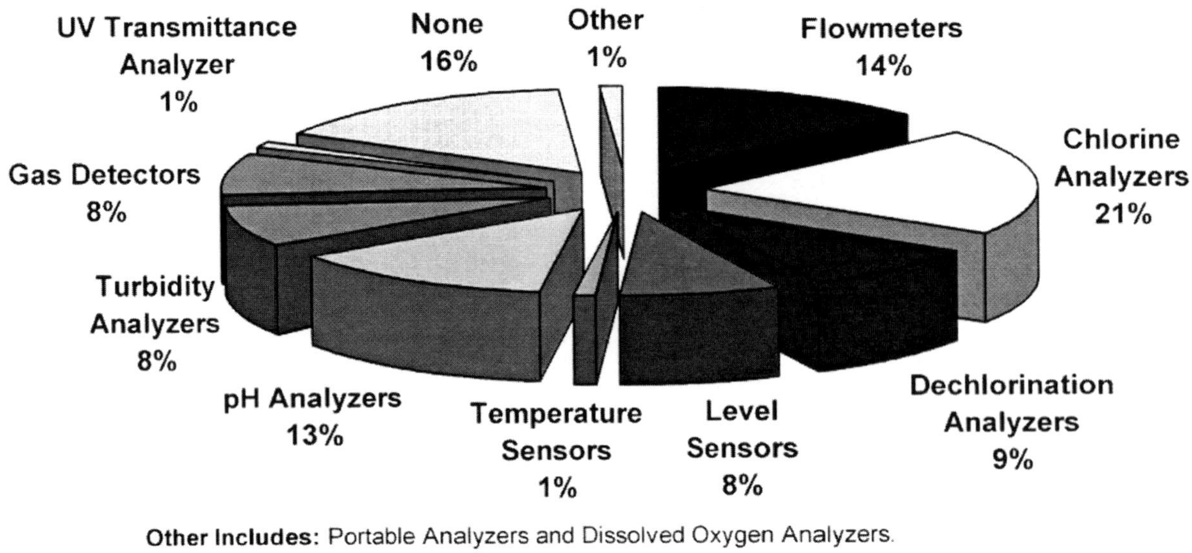

Figure 10-4. Instrumentation.

Figure 10-5 shows that dual power sources (28%), DCS/SCADA systems (22%), multiple control strategies (22%), and portable or in-place generators (16%) are the primary redundant automation and control units used for disinfection and effluent treatment. 3% of treatment plants reported using backup dual field mounted control boxes, and 2% reported using dual PLCs are reported by 2%. "Other" redundant automation and control configurations (3%) include taking one train out of service during seasonal low flows and uninterruptible power supplies. Only 4% of the surveyed treatment plants reported having no redundant automation and control units for their disinfection and effluent treatment processes.

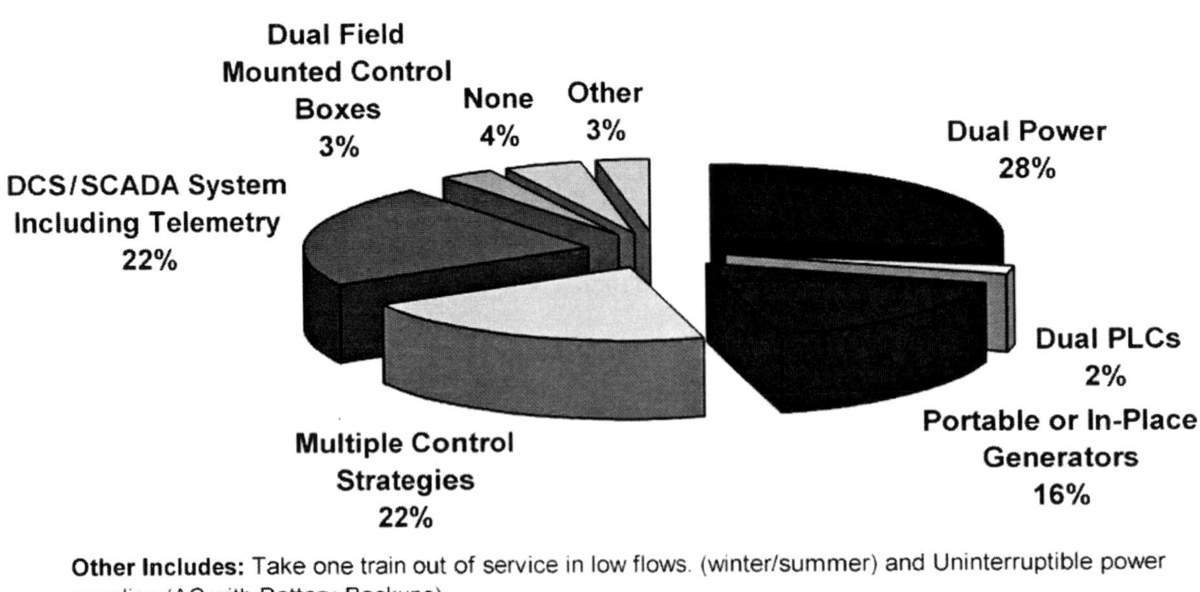

Figure 10-5. Automation and control.

10.4 Redundancy Reasons and Ratings

Figure 10-6 shows that the most frequent reason for implementation of redundancy for disinfection and effluent was reported as in-house operations and maintenance requirements (39%). The second highest reason reported was discharge permit regulations (26%), followed by emergency/disaster planning regulations/requirements (16%) and safety regulations (10%). "Other" reasons (6%) for redundancy include the Water Resources Act, good design practices, and state standards. Only 1% of facilities reported the IEEE Standard 493 as one of the reasons for implementing redundancy for disinfection. 2% reported no redundancy.

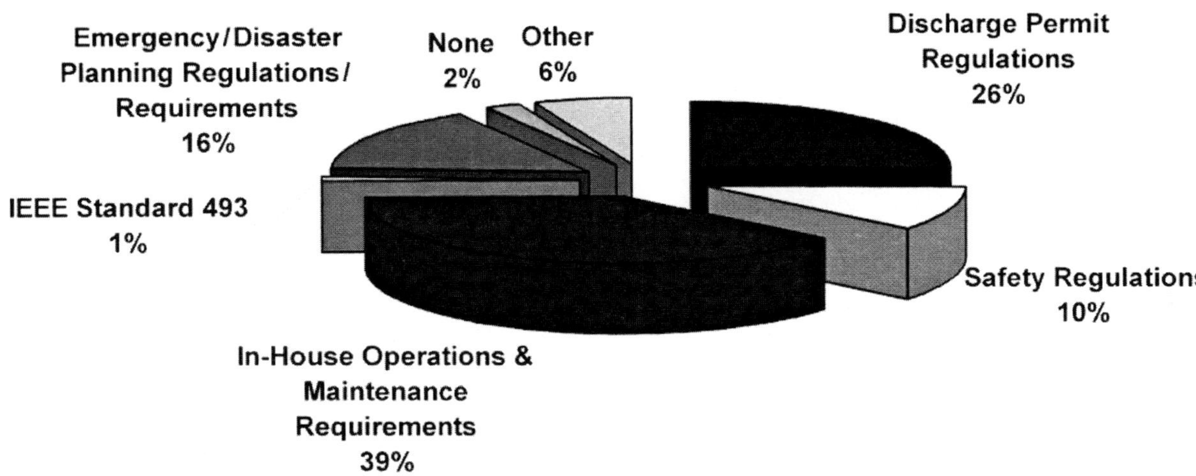

Other Includes: Water Resources Act; Good Design Practices; Federal, State and 10 State Standards for Plant Power Redundancy; and State Standards.

Figure 10-6. Reasons for redundancy.

Figure 10-7 shows that the majority of the treatment facilities that practice redundancy design noted that their redundancy design practices are effective. More than half (55%) reported their redundancy as extremely effective, and 39% reported that their redundancy is somewhat effective. 3% reported that redundancy design practices for disinfection and effluent treatment processes are not necessary, and 3% reported that they never are used.

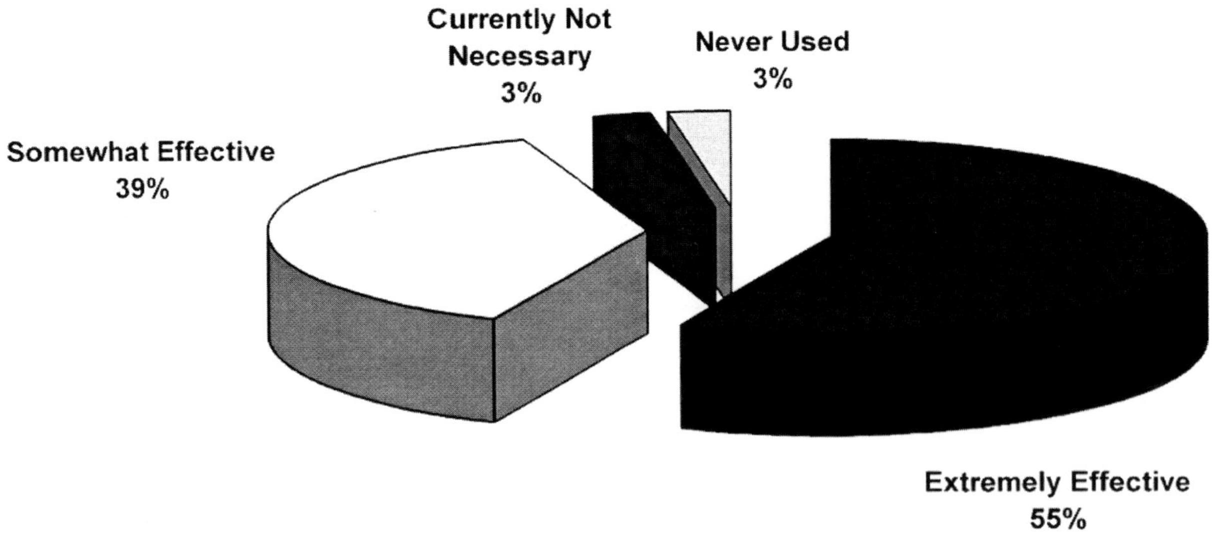

Figure 10-7. Redundancy ratings.

CHAPTER 11.0

Regulatory Overview

11.1 Introduction and Background

The United States federal government enacted the majority of environmental regulations for each state, and the Canadian federal government initiated the major environmental regulations for each province.

The U.S. Congress passed the Clean Water Act (CWA) in 1972, with amendments enacted in 1977 and 1987. These regulations provide environmental standards for WWTP effluent through the NPDES permit program and promote state and local jurisdiction for overseeing the implementation of the CWA regulations.

Other federally enacted legislation that also effects WWTP operations includes the RCRA for the disposal of hazardous wastes and the Marine Protection, Research, and Sanctuaries Act that regulates dumping into ocean waters.

In addition to the federal and state responsibilities for environmental regulations, interstate agencies or commissions were formed to establish guidelines and interactions with other states that share a common interest in protecting the quality of shared waters through environmental standards. These interstate agencies include:

- Great Lakes Commission
- Interstate Commission on the Potomac River Basin
- New England Interstate Water Pollution Control Commission
- Ohio River Valley Water Commission
- Susquehanna River Basin Commission

Federal, state, and interstate agency regulations and guidelines are designed to provide a level of water quality to protect human health and safety through the protection of the environment. Some of these regulations and/or guidelines set effluent limits for WWTPs but do not specifically make provisions for design standards that include redundancy. Instead, the individual states have the jurisdiction in some cases to establish redundancy design criteria to meet federal and state effluent quality requirements.

In Canada, the federal government has given the responsibility of regulating water supply and wastewater treatment discharges to each province, which has established regulations or guidelines—which may or may not incorporate redundancy design criteria in their effluent quality requirements.

This project included interviews with the state regulators to obtain a more informed picture of the redundancy design practices.

11.2 Regulator Survey

Regulatory compliance is thought to be one of the primary driving forces for redundancy design at publicly owned WWTPs. In order to obtain an understanding of the regulatory agency position with respect to redundancy design practices at WWTPs, the project team interviewed U.S. state and Canadian provincial regulators.

The project team requested that each state and province regulator reply to the following seven questions regarding redundancy design practices:

1. Are there redundancy design practice specifications for wastewater treatment facilities in your state/provincial regulations?
   a) If yes, please site the specific regulations (e.g., name and rule number):
   b) How can treatment facilities access this information (via website, etc.)?
2. If there are no redundancy regulations, are there state/provincial redundancy design practice specifications in guidelines or recommended procedures for wastewater treatment facilities?
   a) If yes, please site the guideline or recommendation reference name:
   b) How can treatment facilities access this information (via website, etc.)?
3. If there are no redundancy regulations, guidelines, or recommended procedures, are there plans to implement them in the future?
   a) If yes, when and in what format?
   b) If no, why not?
4. What procedure does your state/province use to implement or enforce the redundancy regulations, guidelines, or recommended practices?
5. Why were redundancy regulations, guidelines, or recommended practices implemented?
   a) For regulatory compliance.
   b) To ensure consistent plant performance.
   c) For protection of sensitive receiving waters.
   d) Other reason, please specify.
6. Do you know if there are any regional or local regulations for redundancy design practices at wastewater treatment facilities?
   a) If yes, please site the specific regulations (e.g., name and rule number):
   b) How can treatment facilities access this information (via website, etc.)?

7. If there are no redundancy regulations at the regional or local level, do you know if there are any guidelines or recommended practices for redundancy design practices for wastewater treatment facilities at the regional or local level?
   a) If yes, please site the guideline or recommended practice reference name.
   b) How can treatment facilities access this information (via website, etc.)?

The following graphs and tables display regulator responses to the survey. In addition, Appendix M provides more detailed information including contact information for state and provincial regulatory agencies, and Appendix N provides supplemental information, including examples of regulations and guidelines. Specifically, Tables M-1 and M-2 provide contact information for states and provincial regulators interviewed for this project, respectively.

The project team contacted regulators from all 50 states and the District of Columbia (Washington, D.C.) as well as all 10 provinces. Only two state regulators (Idaho and West Virginia) did not respond to the survey.

11.3 State/Provincial Redundancy Regulations and Guidelines

Figure 11-1 displays that approximately 41% of the states have regulations with redundancy requirements and approximately 59% of the states do not. Table M-3 in notes that a few of the states have adopted the EPA technical bulletin, "Design Criteria for Mechanical, Electric, and Fluid System and Component Reliability (EPA–430-99-74-001)," for their redundancy design regulations. The remaining states have implemented their own redundancy design regulations. Most of these states provide access to their redundancy regulations via the Internet.

Figure 11-2 shows that the majority of provinces (80%) do not have regulations with redundancy design criteria. 20% reported having provincial regulations with redundancy design requirements for WWTPs.

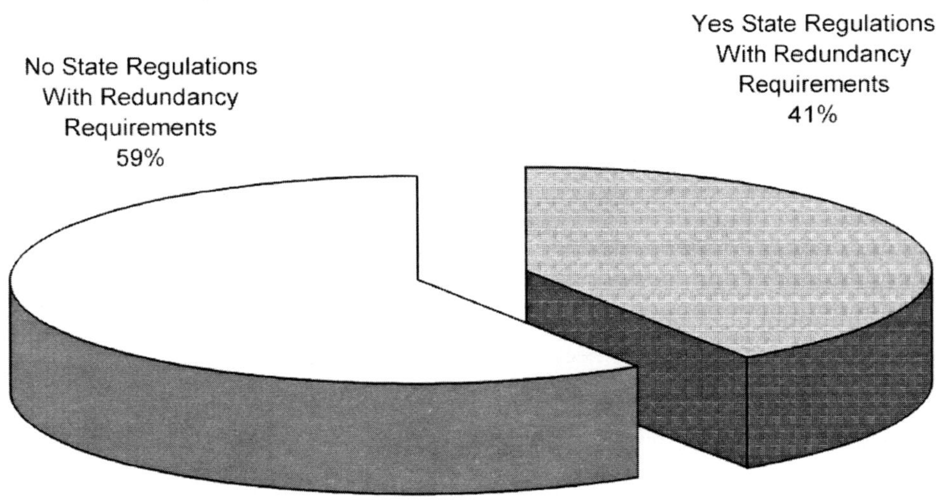

Figure 11-1. U.S. state redundancy regulations.

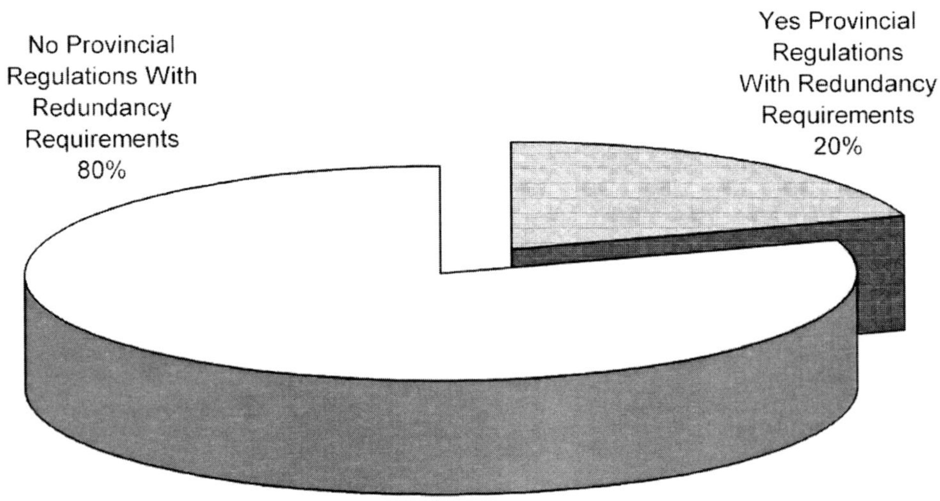

Figure 11-2. Canadian provincial redundancy regulations.

In comparison to redundancy regulations reported by approximately half of the states in Figure 11-1, Figure 11-3 shows that 71% states reported using guidelines for redundancy design practice requirements. Three states have redundancy design regulations and redundancy design guidelines for WWTPs. Approximately 10 states have adopted the 1997 edition of the "Recommended Standards for Wastewater Facilities" also known as the "10-States Standards" (available by hardcopy or Internet from the New York State Health Education Services in Albany, New York). Other states developed their own redundancy design standards or use a variety of references for their guidelines, such as WEF's MOP 8, Metcalf and Eddy's wastewater engineering book, and the EPA's "Design Criteria for Mechanical, Electric, and Fluid System and Component Reliability (EPA–430-99-74-001)."

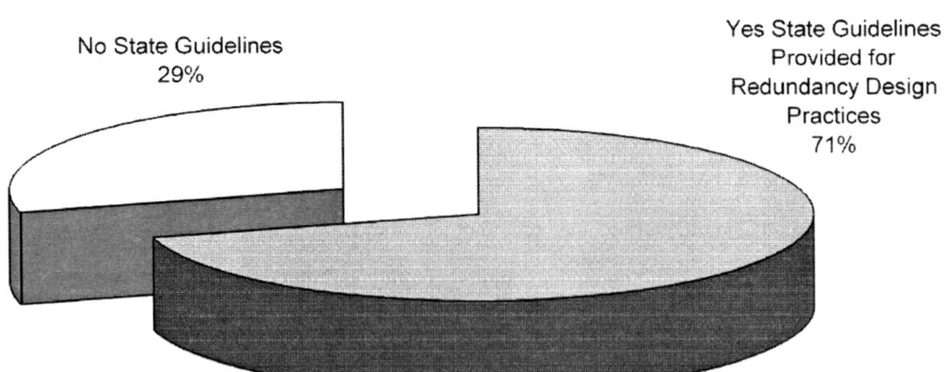

Figure 11-3. U.S. state recommended standards/guidelines.

Figure 11-4 shows that 57% of the provinces use guidelines for redundancy design requirements of WWTPs, while 43% reported no provincial guidelines. Table M-4 notes that the provinces that reported guidelines for WWTPs implemented their own standards. In addition, Table M-4 documents that the provinces with no redundancy design practice guidelines noted that redundancy is implemented through each province's review of a WWTP's design drawings and reports.

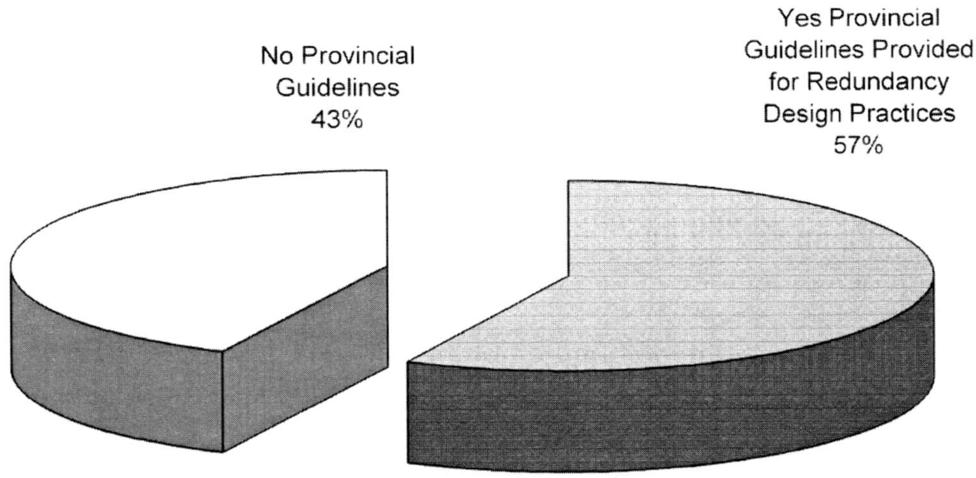

Figure 11-4. Canadian provincial recommended standards/guidelines.

11.4 Local/Regional Redundancy Regulations and Guidelines

A majority of the states (82%) reported that they were not aware of any local or regional redundancy design regulations (Figure 11-5). 18% of states that reported local or regional redundancy design regulations noted that the regional or local authorities adopted the state's regulations (Table M-5). Other states noted using the New England Interstate Environmental Training Center, Guides for the Design of Municipal Treatment Works also known as the "TR-16" document, which has been adopted by most of the New England States. Still more states reported using EPA's "Design Criteria for Mechanical, Electric, and Fluid System and Component Reliability (EPA–430-99-74-001)," and the "10-States Standards."

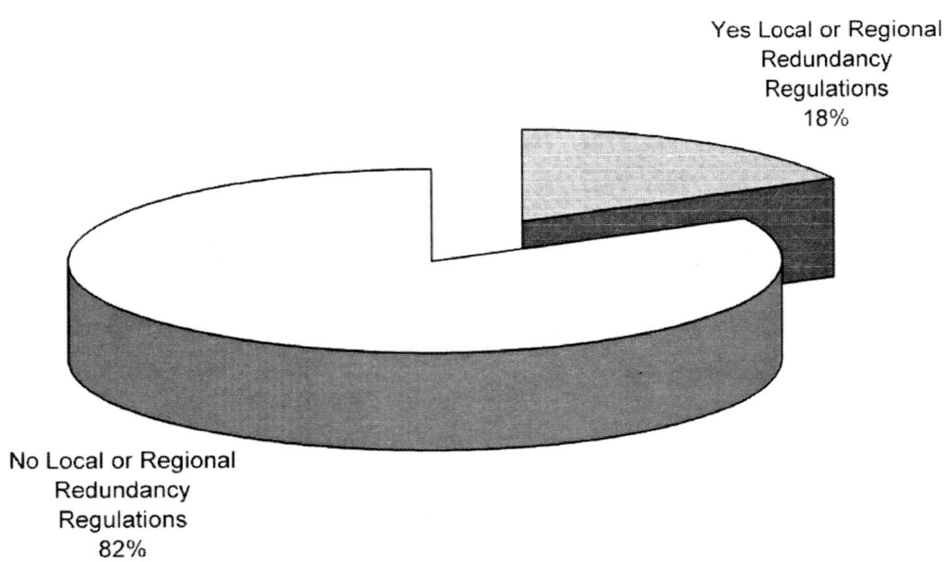

Figure 11-5. U.S. state local/regional regulations.

Figure 11-6 shows that none of the provinces reported having local or regional authorities that have regulations with specific redundancy design criteria.

A majority of states (88%) reported having no knowledge of local or regional guidelines or standards for redundancy design criteria (Figure 11-7). The 12% of the states that did report local or regional guidelines or standards for redundancy design criteria did not know the specific guideline/standard designation but did make reference to New England's TR-16 document and the 10-State Standards.

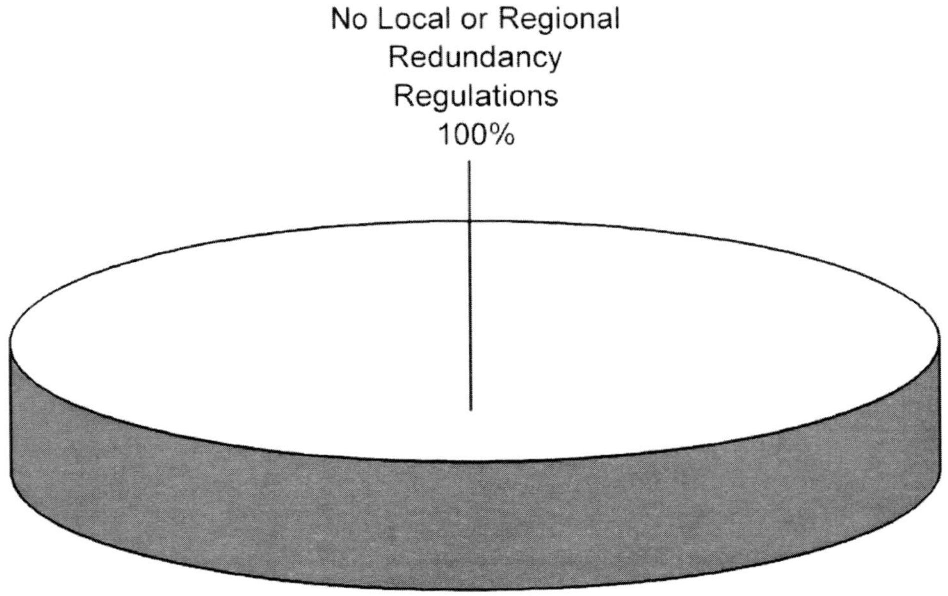

Figure 11-6. Canadian provincial local/regional regulations.

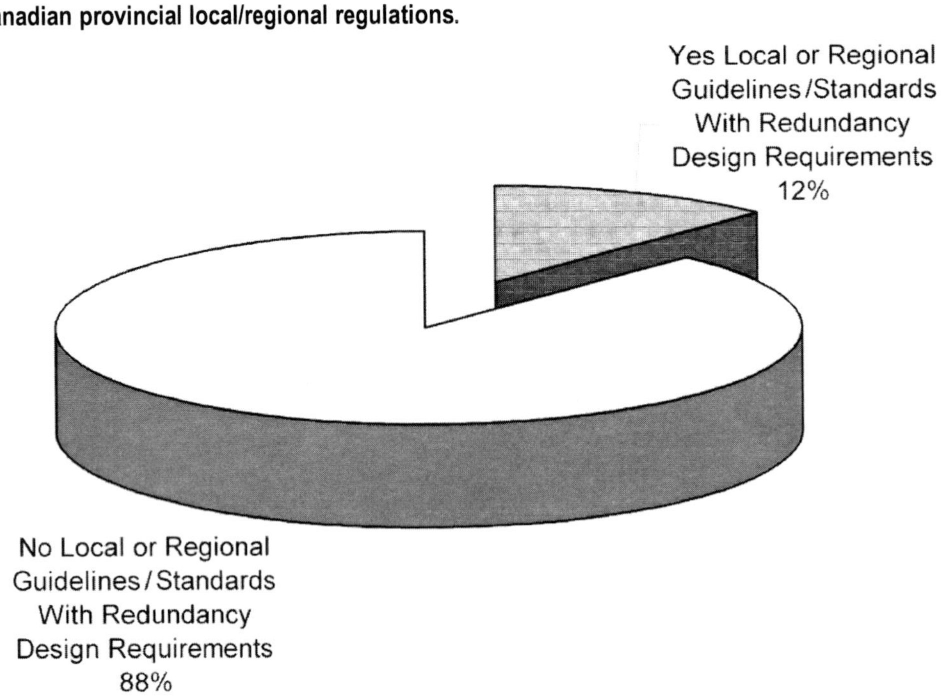

Figure 11-7. U.S. state local/regional guidelines/standards.

Figure 11-8 shows that 13% of the provinces reported having local/regional redundancy design guidelines, while 87% did not.

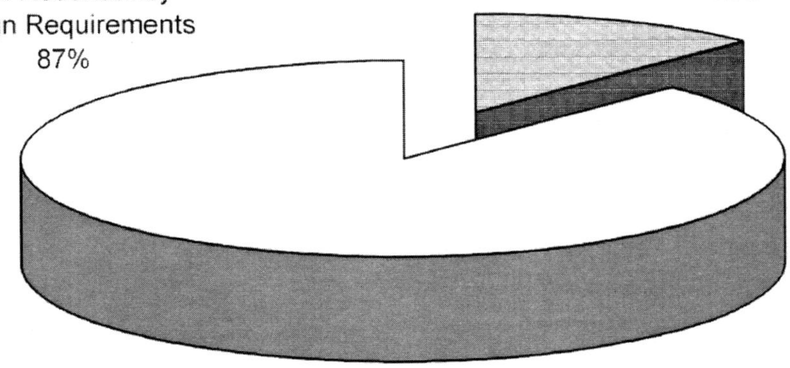

Figure 11-8. Canadian provincial local/regional guidelines/standards.

11.5 Procedures Used to Enforce Redundancy Design Practices in Regulations or Recommended Guidelines

11.5.1 U.S. States

For the United States, the majority of the regulatory agencies reported that construction permits and review of plans and specifications for all WWTP projects (i.e., upgrades and new construction) are used to enforce redundancy design practices. Other facilities reported using operations and maintenance inspections and reports and permit performance histories. Table M-3 provides detailed enforcement responses by the state regulators.

11.5.2 Canadian Provinces

The provincial regulators also reported that enforcement of redundancy design criteria is employed through construction permits and review of plans and specifications for WWTPs. Table M-4 provides detailed enforcement policies as reported by provincial regulators.

11.6 Reasons for Implementing Regulations or Recommended Guidelines

The following provides the reasons for implementing redundancy regulations or guidelines as reported by state regulators. Figure 11-9 shows that each regulator put equal importance to all three selections (regulatory compliance, protection of sensitive receiving waters, and consistent plant performance) as the reasons for implementing redundancy design practices. In addition, 18% of the surveyed regulators reported other reasons that their state implemented redundancy design practices in either regulations or guidelines.

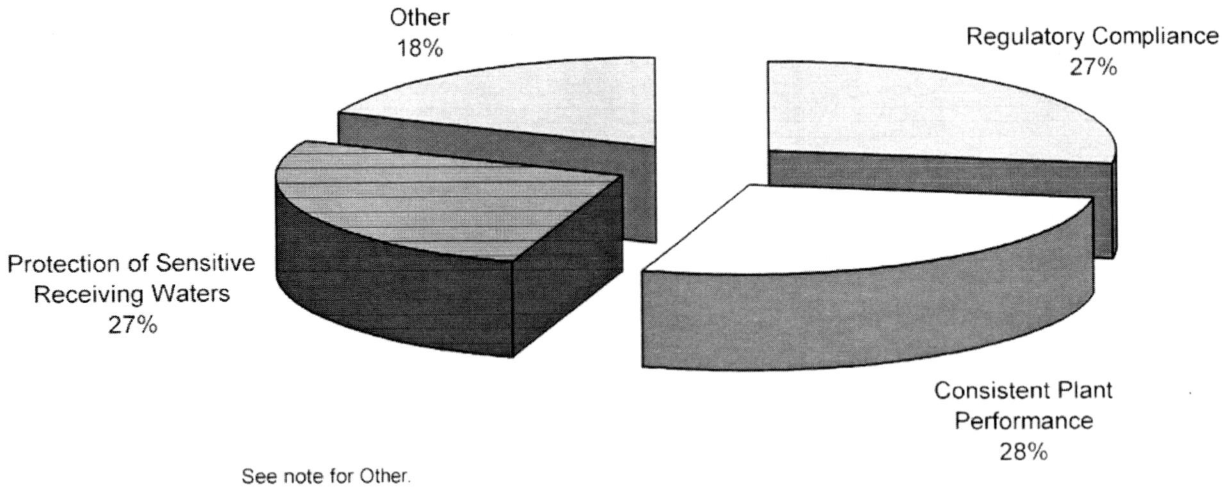

Figure 11-9. U.S. state reasons for implementing redundancy regulations or recommended guidelines.

Other reasons for implementing redundancy design practices reported by state regulators include:

- A provision enacting sanitary improvement district permits for protection of publicly owned treatment works.
- To ensure consistent plant performance and to protect sensitive receiving waters.
- From Delaware statute referenced in Item 4 § 6003. Permit—Required.
  "a) No person shall, without first having obtained a permit from the Secretary, undertake any activity: 1) In a way which may cause or contribute to the discharge of an air contaminant; or 2) In a way which may cause or contribute to discharge of a pollutant into any surface or ground water; or 3) In a way which may cause or contribute to withdrawal of ground water or surface water or both."
- Redundancy guidelines were established to assure treatment plants and collections systems would function in a more reliable manner than was previously accepted. Standby power, dual units, etc., have become common practice today as the public and regulatory authorities expect high compliance rates and consider breakdowns unacceptable. While a higher standard was expected, it had to be better defined technically, and redundancy guidelines aid in that definition. Another need is trying to level the playing field for acceptable equipment. Some suppliers and designers will find a very inexpensive system, which while capable of performing, is not durable and provides no redundancy, but it is less expensive. Sensitive locations are another reason for redundancy or careful design.
- The main purposes of implementing redundancy are to: ensure consistent plant performance so as to meet NPDES permit limits and requirements; protect all waters in the state for beneficial water use; and safeguard public health.
- Public health protection.
- Nebraska looks for enough redundancy to keep collection systems and treatment facilities operating during normal maintenance work and during power outages, including regulatory compliance, consistent plant performance, and protection of receiving waters.

- To ensure operation of each plant's critical processes.
- New Hampshire Department of Environmental Services first developed and imposed WWTP design standards (including process and equipment redundancy requirements) decades ago, probably for the reasons cited above and in keeping with evolving design practice. The design standards were formally codified under state rulemaking procedures in 1975 and have evolved since that time. New Hampshire was the first state to receive EPA construction grants delegation authority in partial recognition of its strong design requirements.
- Redundancy requirements are implemented to protect the public health and receiving water quality.
- For regulatory compliance. Plants were established in mid 1970s, early in EPA construction grants program, and the design had to comply to qualify for grant funds. NPDES permit language supports continuation of fail-safe reliable design to ensure consistent plant performance. Consistent and reliable fail-safe, continuous wastewater treatment state-wide is mandatory per NPDES permitting for the protection of sensitive receiving waters. The main reason for the NPDES program was protection of all receiving waters of the state, regardless of sensitivity. Each NPDES permit establishes and enforces a degree of treatment consistent with the sensitivity of each stream.
- To ensure consistent plant performance and thereby protect water quality.
- For regulatory compliance. To establish minimum criteria for design on some items. To ensure consistent plant performance and for protection of sensitive receiving waters.
- Protection of source water area (drinking water).
- Redundancy requirements are in place for one primary reason: to ensure compliance with the permits that are issued under the provisions of the Tennessee Water Quality Control Act. Secondarily, redundancy requirements provide flexibility in plant operations.
- The Texas Natural Resource Conservation Commission regulations have been in place for close to 50 years. The state design criteria requires some units to be redundant.
- Including regulatory compliance, consistent plant performance and protection of receiving waters, plus it makes a good business sense to stay within compliance and protect environment as a responsible and responsive citizen—corporate, institution or individual.
- Originally, the reliability/redundancy requirements were implemented for federal grant projects. When the Federal Water Quality Act was passed by Congress and was later replaced by the Federal Water Pollution Control Act, large amounts of federal funds were made available to build wastewater treatment facilities. With the funds came many federal regulations for using those funds. The reliability/redundancy criteria were federal requirements. Later, Washington adopted the federal requirements virtually in total.

Figure 11-10 shows the provincial reasons for implementing redundancy. The majority of the provinces (35%) reported other reasons for implementing redundancy design practices as:

- Redundancy is required in situations where public health or environmental risk warrants.
- To ensure that repairs could be undertaken without having to bypass the whole facility.
- To ensure consistent plant performance and to protect sensitive receiving waters
- Regulatory compliance is based on achieving acceptable effluent criteria.
- To provide consistency in the design approach used by process consultants.
- Implemented to protect public health and the environment on a site specific need.

12% of the provincial regulators reported that regulatory compliance was a reason for implementing redundancy design practices, and consistent plant performance and protection of sensitive receiving waters are reported by 29% and 24%, respectively (Figure 11-10).

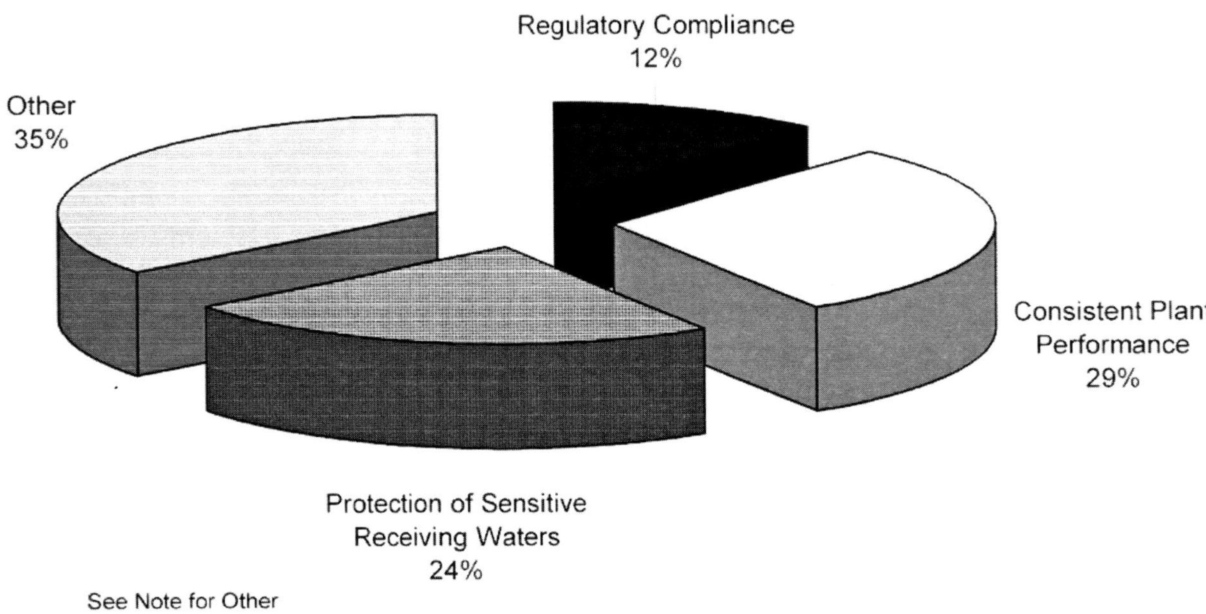

Figure 11-10. Canadian provincial reasons for implementing redundancy regulations or recommended guidelines.

11.7 Redundancy Regulations and Guidelines Summary

Table 11-1 summarizes state and provincial survey responses received regarding redundancy design regulations, and/or guidelines at the state/provincial and regional/local level. In addition, Figures 11-11 and 11-12 display redundancy design regulations and guidelines for states and provinces, respectively.

Table 11-1: U.S. State and Canadian Provincial Redundancy Regulations or Recommended Guidelines/Standards.

State/Province	State/ Provincial Regulations	State/Provincial Recommended Guidelines/ Standards	Regional/Local Regulations	Regional/Local Recommended Guidelines/ Standards
UNITED STATES				
Alabama	Yes	No	Yes	No
Alaska	No	Yes	No	No
Arizona	Yes	No	No	No
Arkansas	No	Yes	No	No
California	Yes			
Colorado	No	Yes	No	No
Connecticut	No	Yes	No	Yes
Delaware	No	No	Yes	
District of Columbia	No	Yes		
Florida	Yes			
Georgia	No	Yes	No	No
Hawaii	Yes	No	Yes	No
Idaho	No Response Received			
Illinois	Yes		No	No
Indiana	No	Yes	No	No
Iowa	Yes	Yes	No	No
Kansas	No	Yes	No	No
Kentucky	Yes		No	No
Louisiana	Yes			
Maine	No	Yes	No	Yes
Maryland	No	Yes	Yes	
Massachusetts	No	Yes	Yes	No
Michigan	No	Yes	No	No
Minnesota	No	Yes	No	No
Mississippi	No	Yes	No	No
Missouri	No	Yes	No	No
Montana	No	Yes	No	No
Nebraska	No	Yes	No	No
Nevada	Yes	Yes	Yes	Yes
New Hampshire	Yes	No	Yes	No
New Jersey	Yes	Yes	No	No
New Mexico	No	No	No	No
New York	No	Yes	No	No
North Carolina	Yes	No	No	No
North Dakota	No	No	No	No
Ohio	No	Yes		Yes
Oklahoma	Yes	No		Yes
Oregon	No	No	No	No
Pennsylvania	No	Yes	No	No
Rhode Island	No	Yes	No	No
South Carolina	Yes	No	No	No
South Dakota	No	Yes	No	No
Tennessee	No	Yes	No	No

Table 11-1: U.S. State and Canadian Provincial Redundancy Regulations or Recommended Guidelines/Standards. *(continued)*

State/Province	State/ Provincial Regulations	State/Provincial Recommended Guidelines/ Standards	Regional/Local Regulations	Regional/Local Recommended Guidelines/ Standards
UNITED STATES				
Texas	Yes	No		
Utah	Yes	Yes	No	No
Vermont	No	Yes	No	No
Virginia	Yes			
Washington	No	Yes	No	No
West Virginia	No Response Received			
Wisconsin	Yes	N/A	No	No
Wyoming	Yes	N/A	No	No
CANADA				
Alberta	Yes			
British Columbia	Yes		No	No
Manitoba	No	No	No	Yes
New Brunswick	No	Yes	No	No
Newfoundland	No	Yes	No	No
Nova Scotia	No	Yes	No	No
Ontario	No	Yes	No	Yes
Prince Edward Island	No	Yes	No	No
Québec	No	No	No	No
Saskatchewan	No	Yes	No	No

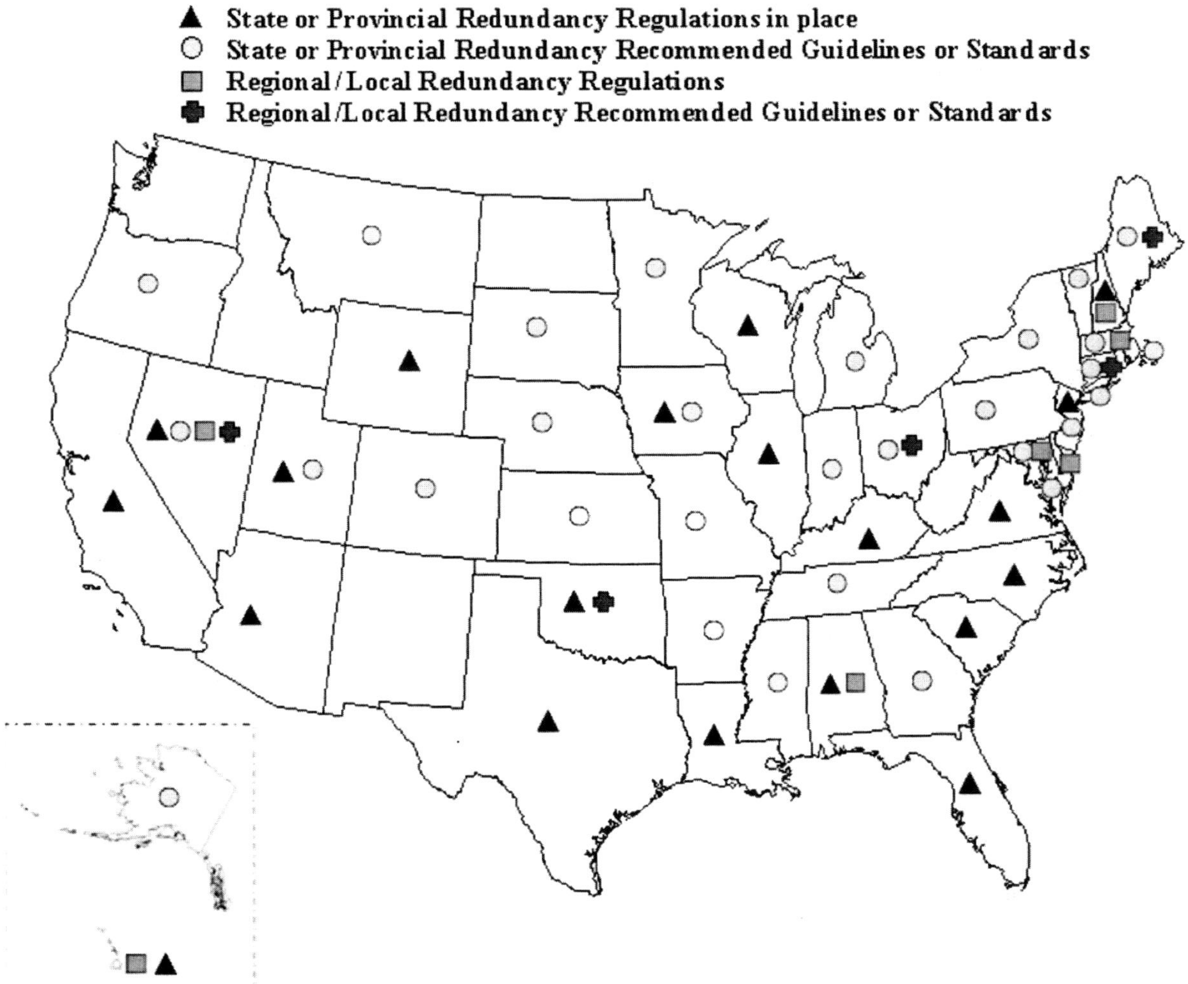

Figure 11-11. U.S. redundancy regulations and guidelines.

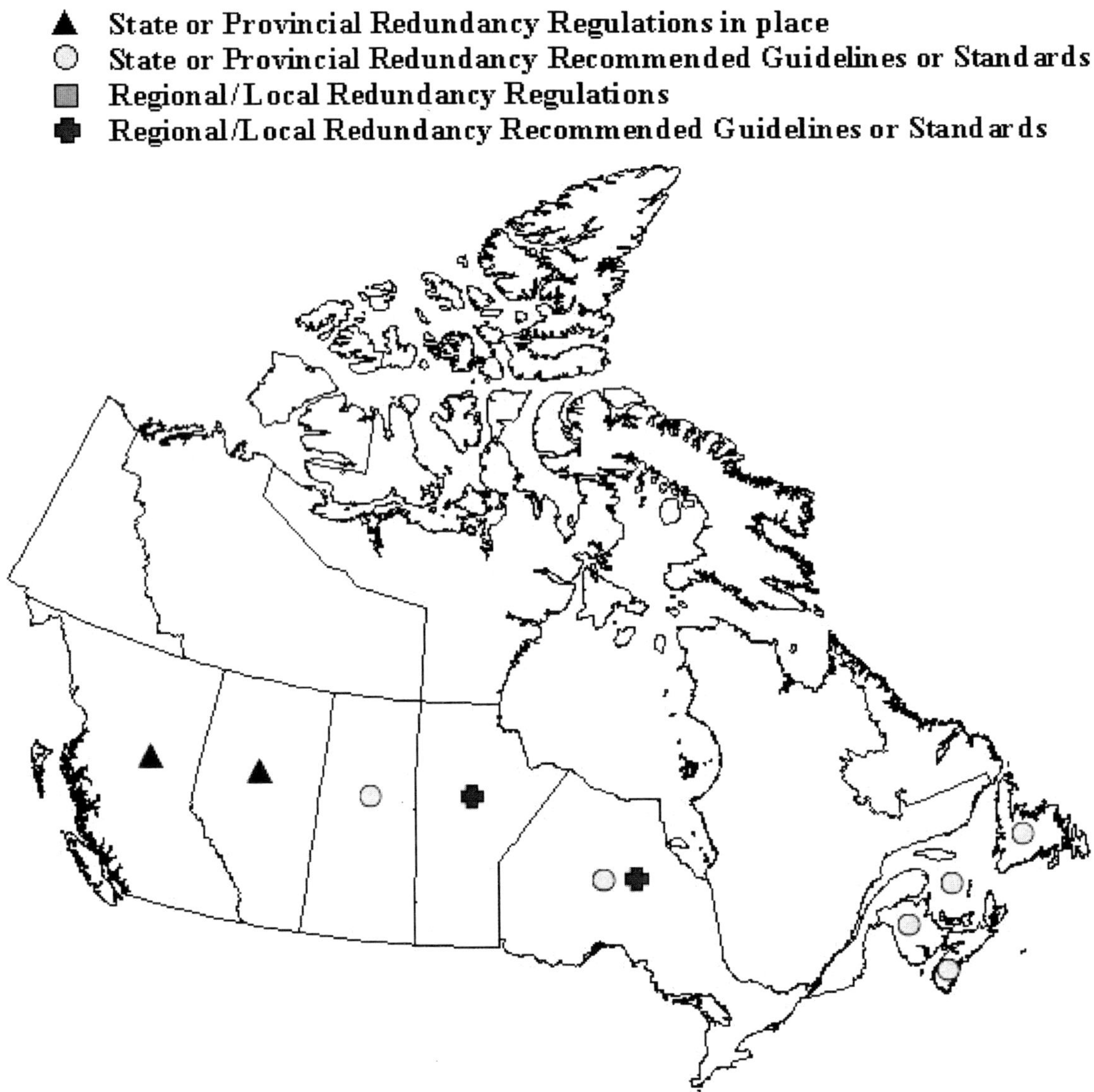

Figure 11-12. Canadian redundancy regulations and guidelines.

11.7.1 Supplemental Information

Appendix N provides supplemental information from the states and provinces regarding regulations and guidelines for redundancy design practices. By far the most impressive supplemental materials are those that provided by Minnesota (Appendix N), which provides its sanitary districts with checklists for the design of specific wastewater treatment processes. Guidelines also are provided on Minnesota's website and can be downloaded using Adobe Acrobat. Minnesota's guidelines provide a good example of ensuring the wastewater effluent quality through the practice of redundancy design. They cover not only treatment process design but also emergency/disaster planning procedures.

In addition, Virginia provided an excellent overview of their states regulations with respect to ensuring redundancy through their reliability provisions. Virginia designates their wastewater treatment plants through a definition of reliability of classes whereby the highest reliability requirements are mandated by a Class I facility as one that is located close to residences or that has a potential discharge into public water supply. Virginia's three classes of reliability in-turn determine the level of redundancy design requirements for a wastewater treatment facility.

CHAPTER 12.0

CASE STUDIES

Case studies are presented to show how system redundancy can be used effectively to increase plant rated capacity through the incorporation of innovative technologies and advanced automated process control. Using the results of the redundancy survey, the project team selected treatment plants that demonstrate success in applying either conventional or alternative redundant design practices. In addition, geographic and treatment facility type played a role in selection. The project team documented successful redundancy design practices as specific case studies by conducting a more in-depth survey of selected facilities through inquiries regarding specific plant practices.

12.1 Introduction and Background

Case studies include data that identify the premise of redundancy design practices and includes redundant design observations made by the following four treatment facilities from the United States and Canada:

- City of Santa Barbara El Estero Wastewater Treatment Plant, Santa Barbara, California
- City of Toronto, Ashbridges Bay Treatment Plant, Toronto, Ontario, Canada
- Orange County Utilities, South Water Reclamation Facility, Orlando, Florida
- Western Lake Superior Sanitary District, Main Treatment Plant, Duluth, Minnesota

Information regarding these plants' redundancy design practices assisted the project team in documenting examples and alternatives to current redundancy methods and the premise behind and efficacy of these practices.

12.2 City of Santa Barbara, El Estero Wastewater Treatment Plant

The City of Santa Barbara, El Estero Wastewater Treatment plant is located along the coast of Southern California, approximately 90 miles north of Los Angeles, in a mild climate where water shortages are common and water reuse is a necessity.

The El Estero plant is a publicly owned and has attended operation seven days per week, 16 hours per day. Seven operations and maintenance staff run this facility, which treats an average flow of 11.0 mgd and peak flows of 16.0 mgd for a service population of 94,000. The plant's wastewater influent flows consist of municipal with industrial pretreatment that comes from a combined sanitary and stormwater system and flows into an activated sludge system with a chlorination and dechlorination process by means of an automation and control system. The plant's effluent is reused for watering local landscape and it is also discharged into the ocean (Figure 12-1).

Santa Barbara defines its primary driving forces for redundancy as operational efficiency and emergency/disaster planning. In addition, Santa Barbara addresses redundancy through design engineer recommendations and California's regulation SB 709 (2000). Documentation of the city's redundancy design practices is found in the plant's operations and maintenance manuals and in the city's policies and procedures.

The El Estero WWTP makes use of alternative redundancy design practices in its operations practices by:

1. Improving primary treatment process clarifier effluent quality by operating with an additional clarifier (i.e., only six clarifiers are needed for plant flow, by adding an additional clarifier online the effluent quality of the primary clarifiers is improved), and
2. Providing the capability of disinfection in the effluent bypass line with a temporary installation.

12.3 City of Toronto, Ashbridges Bay Treatment Plant

The City of Toronto Ashbridges Bay Treatment Plant rests along the western edge of Lake Ontario (Figure 12-2) in the Canadian province of Ontario, located just north of Buffalo, New York. Toronto's WWTP serves approximately 1.25 million customers and treats an average flow of 818,000 m3/d, which is discharged into Lake Ontario. The discharge impacts numerous beaches and the city's drinking water supply. Peak flows can be up to three times the average for their inlet works, pumping stations, primaries, and disinfection basins.

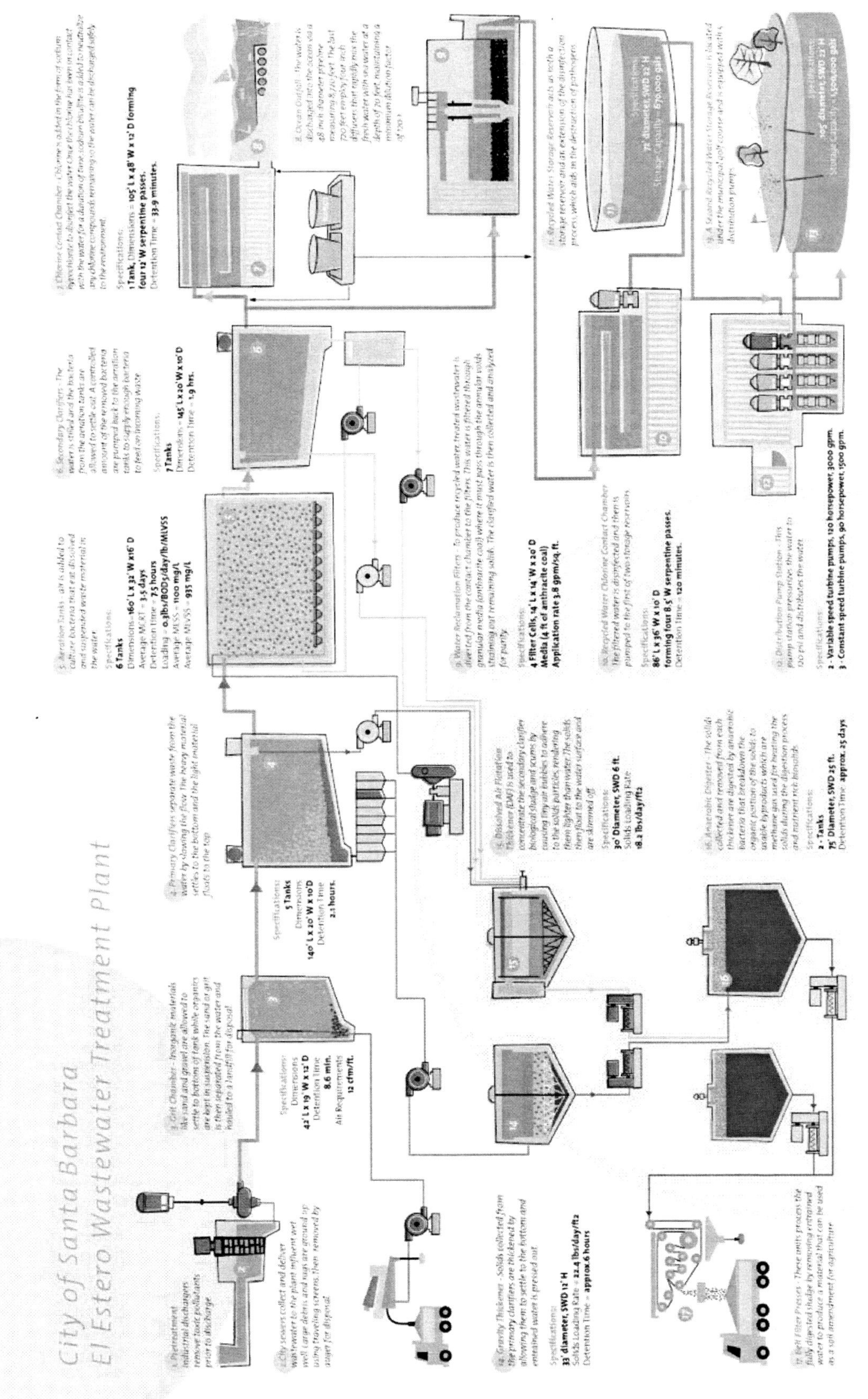

Figure 12-1. El Estero wastewater treatment plant flow schematic.

Figure 12-2. Ashbridges Bay wastewater treatment plant overview.

Ashbridges Bay is a publicly owned and operated WWTP that operates seven days per week, 24 hours per day and is staffed by approximately 200 operations and maintenance personnel. The plant receives its municipal and industrial influent flow from a combined sanitary and stormwater collection system. Toronto's wastewater is treated using suspended growth and chlorination enhanced by the plant's automation and control system.

The plant has redundancy design practices in place for use during peak flows. The primary driving force for Toronto's redundancy design practices is in-house or organizational specifications, policy, and design guidelines. In addition, the city addresses redundancy by using project design team recommendations. Plant personnel documents redundancy design practices in design reports.

12.4 Orange County Utilities, South Water Reclamation Facility

Orange County Utilities' (OCU) South Water Reclamation Facility (SWRF) is located in Orlando, Florida. The SWRF treats an average of 30.5 mgd with a peaking factor of two times the average flow and serves approximately 160,000 people from 50,162+ customer connections.

SWRF is a publicly owned and operated facility that operates seven days per week, 24 hours per day and has approximately 36 operations and maintenance staff. SWRF receives municipal with industrial pretreatment influent flow that is fed from a sanitary collection system and is treated with suspended growth, chlorination, and tertiary filtration. The plant's instrumentation is connected to a plant automation and control system that is used and monitored by plant staff. Final effluent is discharged into percolation ponds and used for reclamation.

OCU has redundant design features in place and uses project design team recommendations to address redundancy issues.

12.5 Western Lake Superior Sanitary District, Main Treatment Plant

The Western Lake Superior Sanitary District's (WLSSD) main treatment plant (Figure 12-3) is located in Duluth, Minnesota, and rests on the southern most tip of Lake Superior approximately 160 miles north of Minneapolis.

Figure 12-3. Western Lake Superior Sanitary District plant overview.

The district's regional treatment plant is designed to treat an average wastewater flow of 43 mgd. The plant is a publicly owned facility that operates seven days per week, 24 hours per day using 40 operations and maintenance personnel.

WLSSD's main treatment facility is an advanced WWTP that receives municipal and industrial flows through a through a 50-mile network of interceptor sanitary sewers. The plant uses an automation and control system that is also connected to a centralized monitoring control system to treat its wastewater.

Wastewater treatment is accomplished through suspended growth using two 40-ton/day oxygen production plants that produce sufficient high-purity oxygen to meet the biological needs of the activated sludge process. In addition, two horizontal liquid oxygen storage tanks help meet peak demand and provide a backup source of liquid oxygen. The plant has four trains with four stages per train and 16 tanks.

Tertiary treatment is accomplished with flocculation and filtration. The activated sludge effluent enters 12 filters, and each contains a layered bed of mixed media: (from top to bottom) medium grain anthracite coal, silica sand, garnet, fine gravel, and medium-size gravel.

Disinfection is accomplished with chlorination and dechlorination, and solids processing is conducted with DAF sludge thickeners and filter belt presses, providing a sludge with a solids content of 16% to 18%. Final wastewater solids are processed in WLSSD's biosolids facility and are either incinerated or lime stabilized for land application. Wastewater effluent is discharged into a recreational receiving stream.

The district's main treatment plant has redundancy design practices in place and uses redundant units during peak flows. The primary driving forces for WLSSD's redundancy design practices include: federal and regional regulations; operational efficiency and maintenance issues; in-house or organizational specifications, policy, and design guidelines; and emergency and disaster planning.

In addition, WLSSD addresses redundancy design through design engineer recommendations and project design team recommendations. These redundancy design practices are documented in: design reports, operation and maintenance manuals, and use of reliability standards that require twice the capacity plus a standby unit that exceeds the 10-States Standards.

12.6 Case Study Questions

The following questionnaire was used to request more in-depth information regarding selected treatment facilities redundancy design practices.

1. Before requesting specific information on the application of redundancy design practices within your organization, we are interested in your opinion on the concepts related to redundancy design practices. The following represent typical issues which are of interest to the study team:
   a) Which design redundancy practices are most effective?
   b) What is the premise for these practices?

c) Why do these design practices work?
d) What level of design redundancy is required for optimal performance?
e) What are the cost benefits of applying redundancy practices?
f) How can these practices be implemented?
g) Can you provide the documentation or a reference to the documentation of your in-house requirements/standards for redundancy design practices?
h) If you do not provide a copy, please identify how to access the document, as we will use it as a reference to our research report if it is available electronically or as a published document.

2. How do your in-house requirements/standards relate to the state or provincial requirements/standards?
3. Is the state or provincial redundancy policy a regulation or a guideline?
4. Was the redundancy policy followed during your last plant expansion?
5. Can you provide a plant process flow schematic and a line drawing of your automation system?
6. Please identify the population served, the average design flow and the design peaking factor.
7. Can you identify process trains, equipment and instrumentation added as a result of the redundancy policy and what impact did each have on the capital cost?
8. What impact has the implementation of the policy had on the following:
   a) System performance
   b) Regulatory compliance
   c) Operating and maintenance costs
   d) Energy utilization
   e) Any other measurable parameters
9. What one factor was of prime importance in your adoption of your current redundancy design practices?
10. What do you consider to be the most important factor in your successful implementation of redundancy design practices?
11. What redundancy design practices that are in use at your utility do you feel are unique or alternative to most common approaches?

12.7 Case Study Survey Results

Appendix N provides a detailed compilation of case study survey questions and answers in a tabular format. The key factors regarding the adoption of redundancy design practices for case study participants are summarized as follows.

12.7.1 Premise of Redundancy Design Practices

OCU reports that the premise for instituting redundancy at their facility is to take advantage of the opportunity to meet regulatory requirements and to reduce operational costs by reducing process unit downtime.

Toronto adds to OCU's statement by reporting that "facilities want to deliver a reliable yet cost-competitive service based on sound preventative maintenance programs." In addition, WLSSD reports that facilities with redundancy process units can afford to schedule preventative maintenance tasks on units not in use with some operational flexibility. Finally, Santa Barbara states that economics dictate a sufficient level of automation and reliable power sources that established the premise for their redundancy programs.

12.7.2 State/Provincial Regulations and/or Guidelines

The province of Ontario provides guidelines for Toronto to implement redundancy practices. A review of design and the issuance of a "certificate of approval" by the province demonstrates that the redundancy guidelines have been met. The province redundancy guidelines regulate standby/backup power requirements and specific capacity provisions of pump stations for prevention of overflows, spills, and basement flooding. In addition to following the provincial guidelines of Ontario, Toronto uses in-house redundancy design practices that closely follow standard design practices prevalent in North America.

For OCU, project designs are approved by Florida state government through the permit process and operation protocols. To meet the state requirements, OCU staff use proven past practices. In addition, OCU staff use new systems designed to meet redundancy requirements that are evaluated by staff and consultants.

WLSSD also has state guidelines that were utilized in the plans and specifications for their recent biosolids project.

12.7.3 Implementation of Redundancy Design Practices

The first crucial step, according to Toronto, for designers and facility staff is to realize that redundancy design practices are site specific and that each treatment process has unique requirements. To further clarify Toronto's statement, redundancy requirements at treatment facilities can vary from the physical treatment plant layout and geographical features to the types and level of treatment processes selected to meet permit requirements. In addition, Toronto noted that it is of prime importance to incorporate management goals and consider staffing levels and experience when implementing redundancy.

Toronto also stresses that implementing successful redundancy design practices relies on a thorough understanding of the treatment process, including redundant process units, equipment, and automation. Toronto further emphasized that the identification by the design team of "critical process elements" is crucial for practical implementation.

Santa Barbara also noted that it is essential for each facility to have reliable treatment in order to meet regulatory and permit limits. Santa Barbara further stated that redundancy within the treatment process at the facility can produce predictable maintenance costs and efficient operations.

WLSSD staff agree with Santa Barbara and state that by implementing redundancy, the treatment process can "produce predictable and efficient operations and maintenance costs."

WLSSD also noted that redundancy is best implemented in original process designs. However, processes that are in-place and operate at a sub-optimal level can also have redundancy added to increase performance. WLSSD further states that facilities managed with a "performance culture could add redundancy to a continuous improvement plan."

Finally, OCU reported that the cost associated with implementing redundancy "directly impacts the owner's ability and willingness to mitigate worse case scenarios" and associated costs.

12.7.4 Redundancy Considerations for Plant Expansions

Case study facilities were asked if their redundancy practices were followed during the last expansion or upgrade. Santa Barbara answered no but noted that with the introduction of state legislation 2000, implementation of redundancy was applied to critical process units. The three other case study participants followed their respective redundancy design practices for expansions and upgrades.

OCU reported that redundancy can also be practiced on processes that do not normally require redundancy and noted that a business case must be made where the additional cost for redundancy can be recaptured in a short term, two to three years, or if the redundancy can reduce staff hours by actual time or by reducing process downtime.

12.7.5 Why Redundancy Design Practices Work

OCU reports that redundancy design practices that work ensure that facility operations and maintenance staff have a viable alternative to reliably meet their regulatory compliance parameters. These working practices also provide assurances that the WLSSD stakeholders needs are met. For example, management attains reliability and efficiency, operations and maintenance gains operational flexibility, and the owners receive economic benefits. The implementation of proven technology, as stated by Toronto, is also a large part of successful redundancy design practices.

12.7.6 Level of Redundancy for Optimal Performance

The level of redundancy design required for optimal performance includes redundancy for critical process components, stated Toronto. This level of redundancy pays careful consideration to the stakeholders needs, noted WLSSD. Optimal redundancy levels at OCU are also dependent upon the current treatment technology, the physical condition of the process unit, and the technical qualifications of facility staff. Optimal performance at OCU also is enhanced by the use of automation computer controlled systems that have redundant or backup features.

Meeting regulatory compliance is an essential performance indicator that facilities are practicing redundancy. OCU has an established record of meeting their compliance parameters to the point that they have received various performance-based awards.

12.7.7 Cost Effective Redundancy Design Practices

The most effective redundancy design practices, according to WLSSD, allow for reduced staffing levels while ensuring reliable backup for the identified critical process units. Toronto states that the level of redundancy required must be sufficient to establish continuous service. This cost-effective level of redundancy will also eliminate or greatly reduce permit violations and provide alternative process control features, according to OCU.

Santa Barbara identified the following process units that make the most effective use of redundancy practices: chlorination and dechlorination monitoring equipment, auxiliary power generators, and influent pumping.

12.7.8 Cost Benefits of Applying Redundancy Design

Toronto reported that the cost benefits of redundancy are realized when the technology used for redundancy is proven to be reliable in design and practice. The redundant process units must also be maintained and ready for use, which incurs additional costs to the facility besides the initial capital costs. Santa Barbara notes that additional cost benefits can be realized in the prevention of monetary fines incurred for permit violations and environmental impacts. OCU states that improvements in operational flexibility and scheduling of maintenance can save overall treatment costs. A conscious decision must be made to balance the costs of installing redundant unit process units and the long-term costs of maintaining the redundant units. Excessive redundancy installations will lead to not only increased capital costs but also escalated long-term maintenance costs.

12.7.9 Demonstrated Factors of Success in Applying Redundancy Design

The commitment from WLSSD staff, management, and owners is reported as one of the most important factors in successful implementation of redundancy design. OCU noted that success comes from experiences in which facilities "strive to prevent the reoccurrence of failure that impacts operational goals" in order to achieve purposeful redundancy design. Another perspective from WLSSD noted that the significant factor in redundancy design is that appropriate steps be taken to ensure that the needs of all stakeholders be considered during the design phase of any project. WLSSD reported that with these factors in place, the redundant processes can help protect public health and the environment within reasonable operating parameters.

12.7.10 Redundancy Design Alternatives

Santa Barbara identified equipment added by their redundancy policy as: chlorine residual analyzers, fixed emergency auxiliary power generators (2.0 megawatts), and a portable power generator (260 kilowatts).

Santa Barbara also stated that this equipment is an alternative approach to most common redundancy design practices.

OCU has a unique approach to cost-effective redundancy. The utility stores enough reclaimed water at their facility to protect the billion-dollar per year citrus industry in central Florida from freeze conditions. OCU reported that the additional storage capacity "directly impacts the local, state, and national economy."

12.7.11 Redundancy Performance Indicators

Regulatory compliance may indicate that facilities are practicing redundancy. OCU facilities have a record of meeting their compliance parameters and have received recognition for their performance standards. WLSSD has no redundancy operational issues concerning compliance except for SSOs in their collection system due to inflow and infiltration.

Santa Barbara's redundancy practices have improved their treatment reliability and ability to meet NPDES standards. Toronto's redundancy program has ensured that their facility complies with Certificate of Approval requirements at all times.

Santa Barbara, Toronto, and OCU stated that redundancy has given their systems performance stability and reliability. OCU also stated that improved system performance due to redundancy requires an increase in their staff's technical and educational abilities.

12.7.12 Impact of Redundancy on Operations and Maintenance Costs

Redundancy has had an impact on operations and maintenance costs at all the case study treatment facilities. WLSSD reports that redundancy is responsible for stable or decreasing user rates over the past three years. The local power company also gave a grant to WLSSD for $141,000 for the net reduction of approximately 700 kW of electrical demand due to their redundancy practices. In addition, WLSSD has experienced a substantial reduction in emergency maintenance.

OCU states that their operating costs have stabilized in the last three years, even though the volume of produced reclaimed water has increased. Their capital improvement project (CIP) costs have also gone up dramatically, but this was expected due to new technology and improvements. Redundancy did not affect energy costs for OCU even though overall energy costs have risen. Finally, the use of redundancy has meant reduced staffing levels for OCU.

Santa Barbara reported that redundancy requirements have initiated major CIPs that will improve system performance. These CIPs will increase sewer rates by 10% for the first year and 3% for the next five years. Santa Barbara also has increased electrical usage by 128% due to additional equipment in-use for redundancy requirements.

Toronto has optimized operating and maintenance costs through the use of redundancy. They also are able to manage electrical demand more effectively.

12.7.13 System Improvements Through Redundancy

Two of the case study participants identified processes that have added redundancy and the impact that their practice had on overall capital expenditures. OCU upgraded their SCADA system to safeguard operations control and to allow data information backup. The critical process units with redundancy at OCU are blowers, pumps, controls, and distributed control units. Additional hydraulic capacity at OCU allows for high or low flow conditions to be controlled for the treatment process.

WLSSD gathered information from a recent biosolids project with a budget of $32.66 million and estimated that an additional 5-10% of project costs were required for redundant equipment, such as "process units, piping, and backup systems." WLSSD staff regard their redundancy design practices as a standard applied to all projects. Staff review design proposals and "challenge designers to justify any designs [that] do not meet the redundancy design practices expected."

Toronto reported that due to the integration of redundancy implementation in projects, it would be difficult to capture redundancy costs associated with construction and equipment.

CHAPTER 13.0

SUMMARY OF PROJECT FINDINGS

13.1 Introduction and Background

The term "redundancy" is used freely, has many different meanings, and, therefore, is not well defined. To assist in understanding redundancy issues and practices, the project team developed a definition of redundancy and its practical applications as it applies to the wastewater treatment process.

This project reveals that equipment redundancy is practiced in the design and operation of water and wastewater treatment plants to improve reliability through the provisions of standby equipment or processes to reduce the risk of failure in order to meet water quality regulations or guidelines. Design guidelines are developed to ensure that critical components retain redundant configurations to eliminate system failures while continuously operating the treatment process (especially during equipment failure and while performing preventative maintenance that requires equipment to be taken out of service).

For the purposes of this project, redundancy was defined as any infrastructure (tanks, etc.), equipment (pumps, motors, etc.), instrumentation (flowmeters, analyzers, etc.), and automation and control (computers, power source, PLCs, etc.) units that are not used during normal operations. This includes backup, secondary, or items used for more than one purpose (multipurpose units). Redundancy also includes installed units (but not spare parts) used for monitoring and control of processes regardless of permit requirements.

Although the definition of redundancy does not include regulatory compliance or reliability standards, the project data shows a direct relationship between treatment process reliability, redundancy design, and regulatory compliance. For example, the regulatory review portion of this project reveals that regulators play a role in ensuring reliability of treatment processes through the approval of treatment plant designs and construction permits as a means to enforce redundancy design regulations or guidelines.

This regulatory approval process is designed to fulfill the need for the facility to meet its discharge parameters through reliable operations. Thus, reliability of the facility's treatment processes as required by regulatory compliance is a function of redundancy design practices.

13.2 Redundancy Design Practice Trends

Equipment redundancy design practices can be used efficiently by applying design standards that implement redundancy only as absolutely necessary for optimum performance, or contradictorily, redundancy design practices can increase capital costs, increase system complexity, and reduce operational efficiency.

13.2.1 Surveyed Plant Profiles

While reviewing the trends of redundancy design practices, it is important to take into consideration the profile of the treatment facilities that participated in this survey. For instance, the majority of the facilities treat wastewater at publicly owned and operated facilities with attended operation seven days per week, 24 hours per day. Most of the surveyed facilities have a sanitary collection system only and receive influent flows consisting of municipal combined with industrial pretreatment; treatment processes are mostly secondary suspended growth; have disinfection is through use of chlorination and dechlorination; and plants operate with automation and control systems.

13.2.2 Prevalent Operational Practices

Operational staffing levels for at least 50% of the surveyed plants reported using more than one operator per 1 mgd of treated wastewater during average flows (Figure 2-2). In addition, survey data analysis revealed that a majority of the treatment plants (65%) operate with less than 50% of rated capacity during average daily flows (Figure 2-3). This indicates that a majority of the plants have more than twice the treatment process capacity for average daily flows and verifies that a majority of the treatment plants have redundancy design practices in place. However, during wet weather flows or peak flows, only 17% of the facilities reported operational capability under 50% of rated capacity. Further analysis of survey data shows that 63% of the facilities reported that redundant units are used to manage peak flow events, while 29% of facilities reported that their redundant units are not in use or not in-place for use during peak flows.

13.2.3 Leading Methods of Design and Construction

Most of the surveyed facilities were built between 1960 and 1985. Figure 2-4 shows that 40 out of 46 facilities reported that their most recent upgrade was completed in the last 10 years. Table 13-1 draws a comparison of design and construction methods of facility upgrades between 1991 and 2000 and current practices in the last two years. This data shows an increase of more than three times the number of facilities using the design-bid-build method versus the traditional methods of outside consultant, in-house staff teams, and separate construction contracts.

Table 13-1. Upgrade Method of Design and Construction.

Year of Upgrade/ Method of Design and Construction	Outside Design Consultant	Team Approach of In-House Staff and Consultant	Construction Contractor Separate from Consultant	Design-Bid-Build Method	In-House Design
1991-2000	37.5%	25%	34.4%	3.1%	0%
2001-2002	29.6%	26.0%	29.6%	11.1%	3.7%

Other factors play a role in the decision making process for redundancy design. Project design teams at 56% of the facilities recommended redundancy designs while 34% of the facilities rely on design engineering recommendations (Figure 2-12).

13.2.4 Primary Driving Forces for Redundancy Design Practices

The comparison of receiving water discharge parameters with the primary driving force of redundancy design practices (Figure 2-11) revealed that regardless of the type of discharge receiving stream (or discharge permit), all treatment facilities reported operational efficiency and maintenance issues as the major driving force behind redundancy design practices.

Table 13-2 also shows that operational and maintenance factors (reported as operational efficiency, maintenance issues, and in-house specifications, policies or design guidelines) is the primary driving force for implementing redundancy design practices, as reported by 49% of the surveyed treatment facilities (Figure 2-10). Regulations were reported as the second highest reason for implementing redundancy design practices (29%). Table 13-2 also notes that 12% reported emergency/disaster planning as the primary driving force for implementing redundancy design practices. Since this survey was conducted prior to the September 11, 2001, terrorist attacks on the United States, the reported percentage for emergency/disaster planning redundancy requirements may increase with the new focus on improving security at the nation's water and wastewater treatment plants. Finally, Table 13-2 shows that only 10% of the surveyed treatment facilities reported using outside recommendations such as consultant and/or equipment supplier recommendations or design guidelines as their primary driving forces.

Table 13-2. Primary Driving Forces for the Origin of Redundancy Design Practices.

Primary Driving Forces	In-House Operations, Maintenance, and Design Factors	Regulations	Emergency/ Disaster Planning	Outside Recommendations
Percentage of Surveyed Treatment Facilities	49%	29%	12%	10%

Table 13-3 adds that operation and maintenance manuals are the primary methods of documenting redundancy design practices (45%). Corresponding to enforcement of redundancy design through the review of design reports by regulators, the second highest method of documenting redundancy design practices is reported as design reports or manuals (38%).

Table 13-3. Documentation of Redundancy Design Practices.

Documentation of Redundancy Design Practices	Operations and Maintenance Manuals	Design Reports or Manuals	No Documentation for Redundancy Practices	Facility Standards (other)
Percentage of Surveyed Facilities	45%	38%	12%	5%

13.2.5 Critical Redundancy Components

Successfully implementation of redundancy depends upon the identification of critical components in the treatment process. Table 13-4 identifies the critical redundancy components by treatment process. The survey data is divided into two major equipment categories: bypass infrastructure and backup equipment and instrumentation, automation, and control.

The data in Table 13-4 shows the current focus of redundancy practices as: pumps, screens, basins and bypasses; dual power sources and generators; level sensors, DO analyzers, chlorine analyzers, pH analyzers, flowmeters; and DCS/SCADA systems and multiple control strategies.

Table 13-4 also shows that the most critical redundant components are:

- Pumping to move flows;
- basins and bypasses for hydraulic capacity;
- level sensors, DO analyzers, and chlorine residual analyzers to detect process changes; and
- DCS/SCADA systems to monitor, control, and automate the treatment process.

Table 13-4. Critical Redundancy Components by Treatment Process.

Treatment Process/ Critical Components	Bypass Infrastructure and Backup Equipment	Instrumentation, Automation and Control
Collection Systems, Lift Stations and Influent Flow	• Bypass discharge to local water body (27%) • Influent storage basin (20%) • Pumps at lift stations (58%)	• Wet well level sensors (37%) • Dual power source (26%) • Portable or in-place generators (25%) • DCS/SCADA systems (19%) • Multiple control strategies (15%)
Industrial Pretreatment	• Flow diversion bypass channel or pipe (19%) • Flow equalization basin (17%) • Influent pumps (20%)	• pH analyzers (16%) • Dual power source (17%) • DCS/SCADA systems (16%) • Portable or in-place generators (16%)
Preliminary Treatment	• Grit removal bypass lines (31%) • Backup screening and grit removal for septage receiving (21%) • Bar screens (32%) • Grit chambers or hoppers (25%)	• Level sensors in bar screen channels (27%) • Dual power source (29%) • DCS/SCADA systems (23%)
Primary Treatment	• Bypass for sedimentation/settling tanks (28%) • Backup Sedimentation/settling tanks (55%) • Sludge pumps (47%) • Scum pumps (36%)	• Water level sensors (20%) • Dual power source (26%) • DCS/SCADA systems (23%)
Secondary Treatment	• Sludge bypass channels or piping (29%) • Suspended growth basins (37%) • RAS/WAS Pumping (26%) • Blowers (23%) • Sludge Pumping (21%)	• Dissolved oxygen analyzers (28%) • Dual power source (28%) • DCS/SCADA systems (25%) • Multiple control strategies (21%)
Tertiary/ Advanced Treatment	• Bypass filtration (42%) • Bypass channels for secondaries (26%) • Filters (26%) • Backwash pumps (22%) • Sludge pumps (20%)	• Dissolved oxygen analyzers (19%) • Level sensors (17%) • Dual power source (26%) • DCS/SCADA systems (23%) • Multiple control strategies (19%)
Solids Processing	• Bypass lines for solids processing (34%) • Digester gas bypass lines (33%)	• Sludge flowmeters (12%) • Chemical flowmeters (12%)

Table 13-4. Critical Redundancy Components by Treatment Process (continued).

Treatment Process/ Critical Components	Bypass Infrastructure and Backup Equipment	Instrumentation, Automation and Control
	• Digesters (17%) • Sludge pumps (28%) • Chemical feed pumps (20%)	• Dual power source (26%) • DCS/SCADA systems (25%)
Disinfection and Effluent	• Disinfection bypass channels/pipes (24%) • Effluent bypass channels/pipes (24%) • Disinfection basins (33%) • Chemical feed pumps (26%) • Chlorinators (22%)	• Chlorine analyzers (21%) • Dual power source (28%) • DCS/SCADA systems (22%) • Multiple control strategies (22%)

13.3 Redundancy Regulatory Trends

Information regarding redundancy design practices in regulations and guidelines was obtained through interviews with state and provincial regulators. Table 13-5 shows that 47% of the interviewed state regulators reported having regulatory requirements for redundancy design practices. In addition, 74% of the interviewed regulators use recommended guidelines or standards for implementing redundancy design criteria. Local and regional regulations and guidelines were reported by 18% and 14%, respectively.

Table 13-5. Redundancy Design Regulations and Guidelines.

Country/ Redundancy Regulations or Guidelines	State/Provincial Regulatory Requirements for Redundancy Design Practices	State/Provincial Guidelines for Redundancy Design Practices	Local/Regional Regulations for Redundancy Design Practices	Local/Regional Guidelines for Redundancy Design Practices
United States	47%	74%	18%	14%
Canada	20%	50%	100%	50%

Some state regulators developed their own redundancy design standards in-house, while others reported adopting a variety of references for their guidelines such as the:

- EPA technical bulletin, "Design Criteria for Mechanical, Electric, and Fluid System and Component Reliability (EPA–430-99-74-001)"
- Interstate agency developed 1997 edition of the "Recommended Standards for Wastewater Facilities" also known as the "10-States Standards"
- Interstate agency developed regional publication "TR-16," used by most of the New England States

U.S. regional or local authorities reported either adopting the state's regulations or the above-listed guidelines.

The majority of the Canadian regulators (50%) reported having recommended redundancy design guidelines or standards in place, while 100% of the regulators noted that local and regional authorities have regulations that require redundancy design practices.

Regulators also reported the reasons their state or province implemented redundancy design regulations or guidelines (Table 13-6). The majority of the regulators who reported "other" as a reason for redundancy design regulations noted protection of receiving waters. This would make "protection of sensitive receiving waters" the primary reason for adopting redundancy design regulations or guidelines for both the United States (45%) and Canada (83%). Other reasons state and provincial regulators adopted redundancy design practices in their regulations or guidelines include consistent plant performance and regulatory compliance.

Table 13-6. Reasons for Implementing Redundancy Design Regulations or Guidelines.

Country/ Reasons	Protection of Sensitive Receiving Waters	Consistent Plant Performance	Regulatory Compliance	Other
United States	28%	28%	27%	17%
Canada	17%	17%	0%	66%

13.4 Successful and Alternative Redundancy Design Practices

Based on site surveys, the best examples of applied conventional and alternative redundancy design practices in municipal and industrial treatment facilities are revealed in the treatment plant survey data and case study information, provided in Chapters 2.0-13.0.

13.4.1 Orange County Utilities

OCU applies redundancy design to their facilities to take advantage of the opportunity to meet regulatory requirements and to reduce operational costs by reducing process unit downtime. OCU staff use past redundancy practices that are proven, yet they also use new systems that are evaluated by in-house staff and consultants to meet regulatory requirements. OCU stated that redundancy can be applied to processes not normally requiring redundancy as long as a financial business case is made and the cost for redundancy can be recaptured in two to three years.

The optimized level of redundancy practiced by OCU depends on current technology, the physical condition of the process unit, and the technical qualifications of facility staff to operate and maintain the level of redundancy selected. This level of redundancy is cost effective at OCU because their operating costs have stabilized in the last three years even though the volume of reclaimed water produced has increased. OCU redundancy levels also resulted in cost savings by the reduction of staffing levels at their facilities. Finally, the critical process units with redundancy at OCU include: aeration blowers; pumps; controls, distributed control units, and SCADA systems; and increased hydraulic capacity for reuse storage.

13.4.2 Western Lake Superior Sanitary District

The commitment from WLSSD staff, management, and owners is reported as one of their most important factors in successful implementation of redundancy design. WLSSD staff regard their redundancy design practices as a standard applied to all projects. Facility staff review design proposals and "challenge designers to justify any designs which do not meet the redundancy design practices expected." WLSSD also noted that redundancy is best implemented in original process designs however, processes that are in-place and operate at a sub-optimal level can also have redundancy added to increase performance. WLSSD further states that facilities managed with a "performance culture could add redundancy to a continuous improvement plan."

WLSSD facilities with redundant process units can afford to schedule preventative maintenance tasks on units not in use with some operational flexibility. WLSSD also has state guidelines that were used in the plans and specifications for their recent biosolids project. WLSSD gathered information from their recent biosolids project with a budget of $32.66 million and estimated that an additional 5-10% of project costs were required for redundant equipment, such as, "process units, piping, and backup systems."

The local power company also gave a grant to WLSSD for $141,000 for the net reduction of approximately 700 kW electrical demand due to their redundancy practices. In addition, WLSSD has experienced a substantial reduction in emergency maintenance.

13.4.3 City of Toronto, Ashbridges Bay Wastewater Treatment Plant

The province of Ontario provides guidelines for Toronto to implement redundancy practices. The first crucial step, according to Toronto, for designers and facility staff is to realize that redundancy design practices are site specific and that each treatment process has unique requirements. Toronto also stresses that implementing successful redundancy design practices depends on a thorough understanding of the treatment process that requires redundant process units, equipment, and automation. Toronto further emphasized that the identification by the design team of "critical process elements" is crucial for practical implementation. In addition, Toronto noted that it is of prime importance to incorporate management goals and consider staffing levels and experience when implementing redundancy.

The implementation of proven technology as stated by Toronto, is also a large part of their successful redundancy design practices. Toronto reported that the cost benefits of redundancy are realized when the technology used for redundancy is proven to be reliable in design and practice. Toronto has optimized operations and maintenance costs through the use of redundancy and also has been able to manage electrical demand more effectively.

13.4.4 City of Santa Barbara, El Estero Wastewater Treatment Plant

Santa Barbara reported that economics dictate a sufficient level of automation and reliable power sources that established the premise for their redundancy programs. Santa Barbara further stated that redundancy within the treatment process at the facility can then produce predictable maintenance costs and efficient operations. Santa Barbara's redundancy practices have improved their treatment reliability and ability to meet NPDES discharge standards. They reported that redundancy requirements have initiated major capital improvement projects that will improve system performance. Santa Barbara noted that additional cost benefits can be realized in the prevention of monetary fines incurred for permit violations and environmental impacts.

Santa Barbara identified the following process units as the most effective use of redundancy practices: chlorination/dechlorination monitoring equipment, auxiliary power generators, and influent pumping.

13.5 Future Research

Additional research is needed to address the many variables that can affect redundancy levels at WWTPs. However, this research project has formulated the many issues that need to be addressed in future planning, design, construction, and regulatory planning of redundancy design practices at WWTPs.

The following list of potential strategic research can be used for identifying future redundancy design research projects:

- Explore the application of redundancy design practices to facilitate emergency, security, and disaster planning issues.
- Research how redundancy is employed during wet weather/peak flow events.
- Conduct further investigation of this project's case study treatment facilities to determine budget planning strategies (including examination of capital costs, operations and maintenance costs, performance of cost/benefit analysis, funding, etc.). These costs can be significant and possibly prohibitive for a utility on a tight budget.
- Accomplish documentation of realistic approachs to redundancy guidelines that have practical goals. Various guidelines can be developed in consideration of high-, medium-, and low-cost options and short- and long-term projects.
- Develop a "how to implement redundancy design practices" manual using this project's case study treatment facilities as a basis.
- Study redundancy issues facing WWTPs based on the required levels of treatment needed to achieve water quality standards.
- Conduct case study research of the operation of expanded and upgraded WWTPs, taking into consideration the size of the treatment plant since there are solutions that can be implemented at smaller plants which do not apply to larger systems and vice versa.
- Examine the effective use of redundant instrumentation and automation, which can have a significant impact on design and operation and study how it can be

used to reduce capital costs or eliminate plant expansions and reduce operations and maintenance costs.

- Create a matrix for experimental designs that would provide the option to cover a wide range of variables and relate them to the need for redundancy in the design of the treatment plant.
- Conduct further research to review current published redundancy guidelines specifically for redundancy practices used by WWTPs and regulators. Redundancy issues for facilities would be addressed based on levels of treatment and these redundancy levels would be related to the regulatory process.
- Develop a redundancy assessment and design tool for use by wastewater treatment designers, facility staff, and regulators. This tool would be based on this project's research and successful case studies.
- Develop a report detailing the pros and cons of specific redundancy practices, including recommendations on the best practices.
- Create a document that details how treatment facilities managed to achieve the level of reliability necessary for their plant's optimal performance.

APPENDIX A

Redundancy Design Practices Survey Form

The following pages display the letter and survey form mailed to ITA members and affiliates. This information was also distributed at WEFTEC 2001 in Atlanta, Georgia. The project team distributed approximately 150 copies.

Instrumentation Testing Association
Promoting Reliability & Performance of Environmental Instrumentation and Automation Technologies

631 N. Stephanie Street, 279
Henderson, NV 89014
USA
Tel: (702) 568-1445
Fax: (702) 568-1446
E-mail: ita@instrument.org
Internet: http://www.instrument.org

October 2001

RE: Research Project 00-CTS-5, Identify Effective Equipment Redundancy Design Practices for Wastewater Treatment Plants

Dear ITA Members and Affiliates:

The Water Environment Research Foundation (WERF) has undertaken the above referenced project. Redundancy design is an issue that is not well defined and can sometimes add a level of complexity and cost not associated with optimum performance and operational efficiency. The goal of this research project is to identify and assess redundancy design practices. We seek your assistance to contribute to this research project by telling us about your current and/or alternative redundancy design practices in place.

Please use the attached WERF Effective Equipment Redundancy Design Practices Survey to provide your valuable input. Survey results and findings will be presented as part of a final report for the project. This report will provide a practical guide for owners, operators, and designers to replicate the documented success for comparable treatment processes regarding equipment redundancy design practices.

In order to keep our research on schedule, we need your completed survey by **November 30, 2001**. Please FAX your survey directly to the Instrument Testing Association (ITA), who is coordinating the survey effort, at (702) 568-1446 or if you would like to respond to this survey electronically, please visit ITA's Web site at: http://www.instrument.org

If you have any questions regarding the survey information requested, please contact ITA at (702) 568-1445. Thank you very much for your cooperation!

Sincerely,

Tony M. Palmer
Principal Investigator
Instrumentation Testing Association

Enclosure: WERF Effective Equipment Redundancy Design Practices Survey

An International Technical and Educational Non-Profit Association

WERF Effective Equipment Redundancy Design Practices Survey

Water Environment Research Foundation (WERF) Project No. 00-CTS-5

Identify Effective Equipment Redundancy Design Practices for Wastewater Treatment Plants

Equipment redundancy is practiced in the design and operation of water and wastewater treatment plants to improve reliability through the provisions of standby equipment or processes to reduce the risk of failure while meeting water quality standards.

Design guidelines are developed to ensure that critical components retain redundant configurations to eliminate system failures while continuously operating the treatment process (especially during equipment failure and while performing preventative maintenance that requires equipment to be taken out of service).

Equipment redundancy design practices can be used efficiently by applying design standards that implement redundancy only as absolutely necessary for optimum performance, or contradictory, redundancy design practices can increase capital costs, increase system complexity, and reduce operational efficiency.

For the purposes of this project, redundancy is defined as any

- **infrastructure *(tanks, etc.),***
- **equipment *(pumps, motors, etc.)***
- **instrumentation *(flowmeters, analyzers, etc.),***
- **automation and control *(computers, power source, PLCs, etc.)***

units that are not used during normal average operations. This includes backup, secondary, or items used for more than one purpose (multipurpose units). Redundancy includes installed units (not spare parts) used for monitoring and control processes regardless of permit requirements.

It is the goal of this project to:

- identify examples of municipal and industrial wastewater treatment plant redundancy methods currently in place,
- identify alternative methods to address treatment plant redundancy,
- identify the premise of these equipment redundancy design practices, and
- to assess the effectiveness of these redundancy design practices.

Participate in this very important project by providing your treatment facility's equipment redundancy planning, design and operational practices. This survey is estimated to take approximately 45-60 minutes to complete.

Check all boxes that apply. Skip treatment processes that do not apply to your plant. *(Note: For organizations with multiple plants, please use one plant that best represents your facility's redundancy design practices).*

Treatment process sections start on pages 3 and 7. For the treatment process redundancy questions, only check the boxes that identify secondary or backup units (e.g., redundant units).

Please take a moment to respond to WERF's Survey by November 30, 2001.
Return survey responses to the ***Instrumentation Testing Association (ITA)*** at 631 N. Stephanie Street #279, Henderson, NV 89014 USA – Phone: 702-568-1445
Or fax survey responses to ITA, Fax: 702-568-1446
Or visit ITA's Web Site http://www.instrument.org to fill-in this survey online.

Treatment Facility Contact Information

(Note: Facility name and location will be included in the final WERF report. Personal contact information will remain confidential and will be used for the purposes of obtaining information regarding survey responses and/or to contact an organization to conduct a more in-depth survey and/or to collect case study information.)

Organization Name		**Treatment Facility Name**	
Contact Person		**Title**	
Phone		**Fax**	
Department		**Email Address**	
Mailing Address		**City**	
State/ Province		**Postal Code**	
Country			

Please Mail for Fax: 702-568-1446 Survey Responses to ITA by November 30, 2001 or Answer This Survey Online at www.instrument.org

General Treatment Facility Information *(please check all that apply)*

Type of Treatment Facility:
☐ Water
☐ Wastewater
☐ Advanced Wastewater *(Tertiary)*
☐ Industrial

Ownership:
☐ Public
☐ Private

Operational Responsibility:
☐ Public
☐ Private
☐ Private (contract operations)

Plant Days of Attended Operation:
☐ 1-3 days per week
☐ 3-6 days per week
☐ 7 days per week

Plant Hours of Attended Operation:
☐ Less than 8 hours per day
☐ 8 hours per day
☐ 10 hours per day
☐ 12 hours per day
☐ 16 hours per day
☐ 24 hours per day

Total Operations and Maintenance Plant Staff:

____

Influent Flow:
☐ Sanitary Sewer Only
☐ Combined System *(sanitary sewer and storm water)*
☐ Other *(please specify)* ______________

Influent Flow Characteristics (Wastewater):
☐ Primarily Industrial
☐ Primarily Municipal
☐ Municipal and Industrial
☐ Municipal w/ Industrial Pretreatment

☐ **Primary Treatment Only**

Secondary:
☐ Attached Growth *(Trickling Filter, RBC, etc.)*
☐ Suspended Growth *(Activated Sludge, etc.)*
☐ Other ______________

Disinfection:
☐ Chlorination *(Chlorine, Hypochlorite, Chlorine Dioxide, etc.)*
☐ Chlorination/Dechlorination
☐ UV
☐ Ozone
☐ None
☐ Other ______________

Advanced:
☐ Filtration
☐ Nutrient Removal *(Nitrogen, Phosphorus, etc.)*
☐ Other ______________

What is your plant's Average Daily Flow in mgd *(please convert m3/d units to mgd)* **?**

____

What is your plant's Wet Weather Flow in mgd *(please convert m3/d units to mgd)***?**

____

What is your plant's Maximum Capacity Design Flow in mgd *(please convert m3/d units to mgd)***?**

____

What year did your treatment facility begin service *(i.e., year of completion of original construction)***?**

____

What is the year of the most recent upgrade completed for your treatment facility?

____

What was the method of design and construction for this upgrade *(check all that apply)?*
☐ In-House Design
☐ Outside *(Consultant)* Design
☐ In-House and Consultant Design
☐ Design-Bid-Build With Single-Source (one outside private company)
☐ Construction Contractor Separate from Consultant
☐ Other (please specify)
____

Is your treatment facility part of a centralized monitoring system for a number of treatment plants?
☐ Yes
☐ No

Is your facility instrumentation connected to a plant control and automation system?
☐ Yes
☐ No

Into what type of receiving waters do you discharge?
☐ Drinking Water Receiving Stream
☐ Recreation Receiving Stream
☐ General Use Receiving Stream (not including drinking water or recreation use)
☐ Percolation Ponds
☐ Reclamation
☐ Ocean
☐ Lake
☐ Other (please specify)
____

During peak flows redundant units are:
☐ Used
☐ Not Used
☐ Not Applicable – Redundant Units are Not In-Place

Does your treatment facility have design redundancy practices in place?
☐ Yes
☐ No
☐ Not Aware

If yes, what is the primary driving force for the origin of these redundancy design practices *(mark all that apply)***?**
☐ Federal Regulations (by permit, etc.)
☐ State/Province Regulations
☐ Regional Regulations *(other than Federal or State/Province, i.e. 10 States Standards)*
☐ Operational Efficiency
☐ Maintenance Issues
☐ Consultant Recommendations / Design Guidelines
☐ Equipment Supplier Recommendations / Design Guidelines
☐ In-House or Organizational Specifications / Policy / Design Guidelines
☐ Emergency / Disaster Planning
☐ Other (please specify) ______________

By what other means does your treatment facility address redundancy?
☐ Design Engineer Recommendations
☐ Project Design Team Recommendation
☐ Contractor Recommendations
☐ Supplier Recommendations
☐ Other (please specify)
____

How are your redundancy design practices documented?
☐ Design Manual
☐ Design Report
☐ O& M Manual
☐ Not Documented
☐ Other (please specify) ______________

Please Mail for Fax: 702-568-1446 Survey Responses to ITA by November 30, 2001 or Answer This Survey Online at www.instrument.org

Water Environment Research Foundation

REDUNDANCY DESIGN PRACTICES BY TREATMENT PROCESS

Please Identify Redundancy Design Practices for Your Treatment Process Areas (check all boxes that apply)

Note: Other treatment Process Areas are Listed on Page 7.

Collection Systems, Lift Stations, and Influent Flow *(please check all redundant design practices that apply)*		***Industrial Pretreatment*** *(please check all redundant design practices that apply)*		***Preliminary (grinding, screening, grit removal, septage, flow equalization, etc.)*** *(please check all redundant design practices that apply)*		***Primary (skimming, sedimentation, etc.)*** *(please check all redundant design practices that apply)*	
Infrastructure		**Infrastructure**		**Infrastructure**		**Infrastructure**	
Bypass Units	**Backup / Secondary or Multipurpose Units**	**Bypass Units**	**Backup / Secondary or Multipurpose Units**	**Bypass Units**	**Backup / Secondary or Multipurpose Units**	**Bypass Units**	**Backup / Secondary or Multipurpose Units**
☐ Bypass Influent Line or Channel into Plant *(either by force main or gravity feed system)* ☐ Bypass lines at lift station(s) ☐ Bypass discharge to local water body (stream, lake, etc.) ☐ None ☐ Other *(please specify)*_______	☐ Influent Storage Basin / Tank *(i.e., for extra storage, bypass influent flow, flow equalization, seasonal storage flow, etc.)* ☐ Collection System Storage Basin / Tank ☐ Lift Station Storage Basin / Tank ☐ Storm Water Storage Basin / Tank ☐ None ☐ Other *(please specify)*_______	☐ Flow Diversion Bypass Channel / Pipe ☐ Influent Flow Channel / Pipeline ☐ Gravity Separation or Sedimentation Bypass ☐ Dissolved Air Flotation (DAF) ☐ Chemical Neutralization or Precipitation Bypass ☐ None ☐ Other *(please specify)*_______	☐ Flow Equalization Basin / Tank ☐ Gravity Separation or Sedimentation Basin / Tank ☐ Chemical Neutralization or Precipitation Basin / Tank ☐ Chemical Feed Tanks ☐ None ☐ Other *(please specify)*_______	☐ Bypass channel for screening units *(besides or in addition to overflow gate / weir or overflow bypass line)* ☐ Grit removal bypass line/channel *(redundant pipeline for conducting routine maintenance and repairing severe piping wear from grit transport)* ☐ Septage Receiving Station Bypass ☐ Bypass Lines for Flow Equalization Basin / Tanks ☐ None ☐ Other *(please specify)*_______	☐ Screening and Grit Removal for Septage Receiving (separate from plant) ☐ Additional Storage for Septage Receiving *(To Hold Potential Industrial Flow that Might Shock or Kill Biological Processes in Plant)* ☐ Flow Equalization Basin / Tanks ☐ None ☐ Other *(please specify)*_______	☐ Bypass for sedimentation / settling tank ☐ None ☐ Other *(please specify)*_______	☐ Sedimentation / Settling Tank ☐ None ☐ Other *(please specify)*_______

Equipment *(Backup / Secondary or Multipurpose Units)*	**Equipment *(Backup / Secondary or Multipurpose Units)***	**Equipment *(Backup / Secondary or Multipurpose Units)***	**Equipment *(Backup / Secondary or Multipurpose Units)***
☐ Pumps at Lift Stations ☐ Portable Pumping ☐ None ☐ Other *(please specify)*_______	☐ Influent Pumps ☐ Screens ☐ Grinders ☐ Chemical Feed Pumps ☐ Sludge Pumps ☐ Skimmers ☐ Mixers ☐ None ☐ Other *(please specify)*_______	☐ Grinders or Comminutors ☐ Bar Screens ☐ Grit Chambers or Hoppers ☐ Aerators in Grit Removal Basin / Tank ☐ Aerators in Septage Receiving Basin / Tank ☐ Septage Receiving Station ☐ Flow Equalization Chambers ☐ Flow Equalization pumps ☐ Odor Control System *(Separate from Plant's Main Odor Control System)* ☐ None ☐ Other *(please specify)*_______	☐ Skimmers ☐ Scum Pumps ☐ Sludge Pumps ☐ None ☐ Other *(please specify)*_______

Please Mail for Fax: 702-568-1446 Survey Responses to ITA by November 30, 2001 or Answer This Survey Online at www.instrument.org

REDUNDANCY DESIGN PRACTICES BY TREATMENT PROCESS

Please Identify Redundancy Design Practices for Your Treatment Process Areas (check all boxes that apply)

Note: Other treatment Process Areas are Listed on Page 7.

Collection Systems, Lift Stations, and Influent Flow *(please check all redundant design practices that apply)*	**Industrial Pretreatment** *(please check all redundant design practices that apply)*	**Preliminary (grinding, screening, grit removal, septage, flow equalization, etc.)** *(please check all redundant design practices that apply)*	**Primary (skimming, sedimentation, etc.)** *(please check all redundant design practices that apply)*
Instrumentation *(Backup / Secondary or Multipurpose Units)*	**Instrumentation *(Backup / Secondary or Multipurpose Units)***	**Instrumentation *(Backup / Secondary or Multipurpose Units)***	**Instrumentation *(Backup / Secondary or Multipurpose Units)***
☐ Influent Flowmeters ☐ Collection Systems Flowmeters ☐ Wet Well Level Sensor ☐ Wet Well Gas Detectors ☐ Storm Water Flowmeters ☐ Accelerometers or Displacement Monitors for Vibration Analysis ☐ None ☐ Other *(please specify)*_______	☐ pH analyzers ☐ Analytical Chemistry Analyzers *(inorganic/organic compounds)* ☐ Flowmeters ☐ Level Sensors ☐ Temperature Sensors ☐ Oil and grease detectors ☐ Coagulant Controller ☐ None ☐ Other *(please specify)*_______	☐ Level sensor in bar screen channels ☐ Level sensor in septage receiving basin/tank ☐ Level sensor in flow equalization basin/tank ☐ Air Flowmeter in grit chambers ☐ pH analyzer in septage receiving station ☐ Flowmeters for flow equalization basin/tank ☐ None ☐ Other *(please specify)*_______	☐ Water Level Sensors ☐ Loss of Flow Sensors ☐ Interface / Sludge Blanket Level ☐ Suspended Solids / Sludge Density Meters ☐ None ☐ Other *(please specify)*_______
Automation and Control *(Backup / Secondary or Multipurpose Units)*	**Automation and Control *(Backup / Secondary or Multipurpose Units)***	**Automation and Control *(Backup / Secondary or Multipurpose Units)***	**Automation and Control *(Backup / Secondary or Multipurpose Units)***
☐ Dual Power Source ☐ Dual PLC's ☐ Portable or In-Place Generators ☐ Multiple Control Strategies ☐ DCS/SCADA System (supervisory control and data acquisition) including telemetry ☐ Dual Field Mounted Control Boxes ☐ None ☐ Other *(please specify)*_______	☐ Dual Power Source ☐ Dual PLC's ☐ Portable or In-Place Generators ☐ Multiple Control Strategies ☐ DCS/SCADA System including telemetry ☐ Dual Field Mounted Control Boxes ☐ None ☐ Other *(please specify)*_______	☐ Dual Power Source ☐ Dual PLC's ☐ Portable or In-Place Generators ☐ Multiple Control Strategies ☐ DCS/SCADA System including telemetry ☐ Dual Field Mounted Control Boxes ☐ None ☐ Other *(please specify)*_______	☐ Dual Power Source ☐ Dual PLC's ☐ Portable or In-Place Generators ☐ Multiple Control Strategies ☐ DCS/SCADA System including telemetry ☐ Dual Field Mounted Control Boxes ☐ None ☐ Other *(please specify)*_______

Please Mail for Fax: 702-568-1446 Survey Responses to ITA by November 30, 2001 or Answer This Survey Online at www.instrument.org

REDUNDANCY DESIGN PRACTICES BY TREATMENT PROCESS

Please Identify Redundancy Design Practices for Your Treatment Process Areas (check all boxes that apply)

Note: Other treatment Process Areas are Listed on Page 7.

Collection Systems, Lift Stations, and Influent Flow *(please check all redundant design practices that apply)*	***Industrial Pretreatment*** *(please check all redundant design practices that apply)*	***Preliminary (grinding, screening, grit removal, septage, flow equalization, etc.)*** *(please check all redundant design practices that apply)*	***Primary (skimming, sedimentation, etc.)*** *(please check all redundant design practices that apply)*
Reasons for Redundancy *(Federal, State/Provincial, or Regional Regulations, etc.)*	**Reasons for Redundancy *(Federal, State/Provincial, or Regional Regulations, etc.)***	**Reasons for Redundancy *(Federal, State/Provincial, or Regional Regulations, etc.)***	**Reasons for Redundancy *(Federal, State/Provincial, or Regional Regulations, etc.)***
☐ **Flood Control District Regulations** ☐ **Inflow / Infiltration (I/I) Regulations** ☐ **Stormwater Regulations** *(State/Provincial/ Regional)* ☐ **Sanitary Sewer Overflow (SSO) Regulations** *(State/Provincial/ Regional)* ☐ **Safety Regulations** *(Occupational Health and Safety Admin.–OSHA; National Fire Protection Assoc –NFPA; Uniform Fire Code –UFC; National Institute of Occupational Safety & Health –NIOSH, etc.)* ☐ **In-House Operations & Maintenance Requirements** ☐ **IEEE Standard 493** *(Institute of Electrical and Electronic Engineers recommended practice for reliable electrical system design)* ☐ **Emergency / Disaster Planning Regulations/ Requirements** ☐ **None** ☐ **Other** *(please specify the name of the regulation that applies)* ____________	☐ **General Pretreatment Regulations** *(State/Provincial /Regional)* ☐ **In-House Pollution Prevention Program (product recovery)** ☐ **Discharge Permit Regulations** *(i.e., NPDES – National Pollutant Discharge Elimination System Permit Requirements)* ☐ **Safety Regulations** *(Occupational Health and Safety Admin.–OSHA; National Fire Protection Assoc –NFPA; Uniform Fire Code –UFC; National Institute of Occupational Safety & Health –NIOSH, etc.)* ☐ **In-House Operations & Maintenance Requirements** ☐ **IEEE Standard 493** *(Institute of Electrical and Electronic Engineers recommended practice for reliable electrical system design)* ☐ **Emergency / Disaster Planning Regulations/ Requirements** ☐ **None** ☐ **Other** *(please specify the name of the regulation that applies)* ____________	☐ **Septage Treatment and Disposal Regulations** ☐ **Safety Regulations** *(Occupational Health and Safety Admin.–OSHA; National Fire Protection Assoc –NFPA; Uniform Fire Code –UFC; National Institute of Occupational Safety & Health –NIOSH, etc.)* ☐ **In-House Operations & Maintenance Requirements** ☐ **IEEE Standard 493** *(Institute of Electrical and Electronic Engineers recommended practice for reliable electrical system design)* ☐ **Emergency / Disaster Planning Regulations/ Requirements** ☐ **None** ☐ **Other** *(please specify the name of the regulation that applies)* ____________	☐ **Discharge Permit Regulations** *(i.e., NPDES – National Pollutant Discharge Elimination System Permit Requirements)* ☐ **Safety Regulations** *(Occupational Health and Safety Admin.–OSHA; National Fire Protection Assoc – NFPA; Uniform Fire Code –UFC; National Institute of Occupational Safety & Health –NIOSH, etc.)* ☐ **In-House Operations & Maintenance Requirements** ☐ **IEEE Standard 493** *(Institute of Electrical and Electronic Engineers recommended practice for reliable electrical system design)* ☐ **Emergency / Disaster Planning Regulations/ Requirements** ☐ **None** ☐ **Other** *(please specify the name of the regulation that applies)* ____________

Please Mail for Fax: 702-568-1446 Survey Responses to ITA by November 30, 2001 or Answer This Survey Online at www.instrument.org

REDUNDANCY DESIGN PRACTICES BY TREATMENT PROCESS

Please Identify Redundancy Design Practices for Your Treatment Process Areas (check all boxes that apply)

Note: Other treatment Process Areas are Listed on Page 7.

Collection Systems, Lift Stations, and Influent Flow *(please check all redundant design practices that apply)*	***Industrial Pretreatment*** *(please check all redundant design practices that apply)*	***Preliminary (grinding, screening, grit removal, septage, flow equalization, etc.)*** *(please check all redundant design practices that apply)*	***Primary (skimming, sedimentation, etc.)*** *(please check all redundant design practices that apply)*
Redundancy Ratings	**Redundancy Ratings**	**Redundancy Ratings**	**Redundancy Ratings**
For the Redundancy Design Practices selected above for this treatment process, please rate the effectiveness of implementing these practices: ☐ **Extremely Effective** – Consistently use redundant systems and have been able to reduce failures with these practices. ☐ **Somewhat Effective** – Occasionally use redundant systems for preventative maintenance not for system failures. ☐ **Currently Not Necessary** – The redundant systems of this treatment process area is not currently used but may be used with future growth requirements. ☐ **Never Used** – The redundant systems have never been used and do not foresee a future use.	***For the Redundancy Design Practices selected above for this treatment process, please rate the effectiveness of implementing these practices:*** ☐ **Extremely Effective** – Consistently use redundant systems and have been able to reduce failures with these practices. ☐ **Somewhat Effective** – Occasionally use redundant systems for preventative maintenance not for system failures. ☐ **Currently Not Necessary** – The redundant systems of this treatment process area is not currently used but may be used with future growth requirements. ☐ **Never Used** – The redundant systems have never been used and do not foresee a future use.	***For the Redundancy Design Practices selected above for this treatment process, please rate the effectiveness of implementing these practices:*** ☐ **Extremely Effective** – Consistently use redundant systems and have been able to reduce failures with these practices. ☐ **Somewhat Effective** – Occasionally use redundant systems for preventative maintenance not for system failures. ☐ **Currently Not Necessary** – The redundant systems of this treatment process area is not currently used but may be used with future growth requirements. ☐ **Never Used** – The redundant systems have never been used and do not foresee a future use.	***For the Redundancy Design Practices selected above for this treatment process, please rate the effectiveness of implementing these practices:*** ☐ **Extremely Effective** – Consistently use redundant systems and have been able to reduce failures with these practices. ☐ **Somewhat Effective** – Occasionally use redundant systems for preventative maintenance not for system failures. ☐ **Currently Not Necessary** – The redundant systems of this treatment process area is not currently used but may be used with future growth requirements. ☐ **Never Used** – The redundant systems have never been used and do not foresee a future use.
Comments	**Comments**	**Comments**	**Comments**
Please provide any general comments and/or identify alternative redundancy design practices. Attach additional sheets for comments if necessary.			

Please Mail for Fax: 702-568-1446 Survey Responses to ITA by November 30, 2001 or Answer This Survey Online at www.instrument.org

WERF Effective Equipment Redundancy Design Practices Survey Project 00-CTS-5 — *Page 7 of 10*

REDUNDANCY DESIGN PRACTICES BY TREATMENT PROCESS

Please Identify Redundancy Design Practices for Your Treatment Process Areas (check all boxes that apply)

Secondary ***(attached growth, suspended growth, etc.)*** *(please check all redundant design practices that apply)*		***Tertiary / Advanced Treatment*** ***(filtration, nutrient removal, etc.)*** *(please check all redundant design practices that apply)*		***Solids Processing*** ***(digestion, thickening, dewatering, composting, etc.)*** *(please check all redundant design practices that apply)*		***Disinfection and Effluent*** ***(chlorination, dechlorination, UV, ozone, etc.)*** *(please check all redundant design practices that apply)*	
Infrastructure		**Infrastructure**		**Infrastructure**		**Infrastructure**	
Bypass Units	**Backup / Secondary or Multipurpose Units**	**Bypass Units**	**Backup / Secondary or Multipurpose Units**	**Bypass Units**	**Backup / Secondary or Multipurpose Units**	**Bypass Units**	**Backup / Secondary or Multipurpose Units**
☐ Sludge Bypass Channels or piping ☐ Oxygen Bypass Line ☐ None ☐ Other *(please specify)*_______	☐ Secondary Process Retention Basin / Tanks ☐ Suspended Growth Basin / Tanks *(i.e., aeration basins)* ☐ Attached Growth Basin / Tanks *(i.e., trickling filters, etc.)* ☐ Oxygen Storage Tanks ☐ None ☐ Other *(please specify)*_______	☐ Bypass channel for secondaries ☐ Bypass filtration ☐ None ☐ Other *(please specify)*_______	☐ Nutrient Removal Basin / Tank ☐ Filters ☐ Backwash Tank ☐ None ☐ Other *(please specify)*_______	☐ Bypass lines for solids processing ☐ Digester Gas Bypass lines ☐ None ☐ Other *(please specify)*_______	☐ Digesters ☐ Centrifuges ☐ Gravity Belt Presses ☐ Gravity Belt Thickeners ☐ Plate Filter Presses ☐ Dissolved Air Flotation (DAF) ☐ Digester Gas Holding Tanks ☐ Digester Gas Flares ☐ Chemical Feed Tanks ☐ None ☐ Other *(please specify)*_______	☐ Disinfection bypass channel / pipes ☐ Effluent bypass channel / pipes ☐ None ☐ Other *(please specify)*_______	☐ Disinfection Basin / Tank ☐ Effluent Holding Tank ☐ Tankage (aeration) ☐ Tankage (final clarifiers) ☐ Tankage (liquid oxygen and vaporization storage) ☐ None ☐ Other *(please specify)* _______
Equipment *(Backup / Secondary or Multipurpose Units)*		**Equipment *(Backup / Secondary or Multipurpose Units)***		**Equipment *(Backup / Secondary or Multipurpose Units)***		**Equipment *(Backup / Secondary or Multipurpose Units)***	
☐ RAS/WAS Pumping *(Return/Waste Activated Sludge)* ☐ Sludge Pumping ☐ Mixers ☐ Aerators ☐ Blowers ☐ Oxygen Compressors ☐ Chlorinators ☐ None ☐ Other *(please specify)*_______		☐ Sludge Pumps ☐ Filter Pumps ☐ Backwash Pumps ☐ Filter Media ☐ Aerators ☐ Baffels ☐ Odor Control System ☐ None ☐ Other *(please specify)*_______		☐ Sludge Pumps ☐ Gas Compressors ☐ Chemical Feed Pumps ☐ Chemical Mixers ☐ Odor Control System ☐ Boilers / Steam Supply ☐ None ☐ Other *(please specify)*_______		☐ Effluent Pumps ☐ Chemical Feed Pumps ☐ Chemical Mixers ☐ UV Lamps ☐ Ozone Generators ☐ Chlorinators ☐ Dechlorinators ☐ Air Scrubbers ☐ None ☐ Other *(please specify)*_______	

Please Mail for Fax: 702-568-1446 Survey Responses to ITA by November 30, 2001 or Answer This Survey Online at www.instrument.org

REDUNDANCY DESIGN PRACTICES BY TREATMENT PROCESS

Please Identify Redundancy Design Practices for Your Treatment Process Areas (check all boxes that apply)

Secondary (attached growth, suspended growth, etc.) *(please check all redundant design practices that apply)*	**Tertiary / Advanced Treatment (filtration, nutrient removal, etc.)** *(please check all redundant design practices that apply)*	**Solids Processing (digestion, thickening, dewatering, composting, etc.)** *(please check all redundant design practices that apply)*	**Disinfection and Effluent (chlorination, dechlorination, UV, ozone, etc.)** *(please check all redundant design practices that apply)*
Instrumentation *(Backup / Secondary or Multipurpose Units)*	**Instrumentation *(Backup / Secondary or Multipurpose Units)***	**Instrumentation *(Backup / Secondary or Multipurpose Units)***	**Instrumentation *(Backup / Secondary or Multipurpose Units)***
☐ Dissolved Oxygen Analyzers ☐ Flowmeters (RAS/WAS) ☐ Air Flowmeter for blowers ☐ Interface / Sludge Blanket Level ☐ Water Level Sensors ☐ Suspended Solids Analyzers ☐ Oxygen Purity Analyzers ☐ None ☐ Other *(please specify)*_______	☐ Level sensors ☐ Ammonia Analyzer ☐ Nitrate/Nitrite Analyzer ☐ Phosphorus Analyzer ☐ pH analyzer ☐ Dissolved Oxygen Analyzer ☐ BOD Analyzer (Biochemical Oxygen Demand) ☐ COD Analyzer (Chemical Oxygen Demand) ☐ Respirometer ☐ Flowmeters ☐ ORP (Oxidation Reduction Potential) Meter ☐ None ☐ Other *(please specify)*_______	☐ Suspended Solids Analyzers ☐ Sludge Flowmeters ☐ Chemical Feed Flowmeters ☐ Gas Flowmeters ☐ Temperature Sensors ☐ Gas Detectors ☐ Fire Sensors ☐ Level Sensors ☐ Solids Volume, Weight and Concentration Meters ☐ None ☐ Other *(please specify)*_______	☐ Flowmeters ☐ Chlorine Analyzers ☐ Dechlorination Analyzers ☐ Ozone Analyzers ☐ Level Sensors ☐ Temperature Sensors ☐ pH Analyzers ☐ ORP Meters (oxidation-reduction potential) ☐ Turbidity Analyzers ☐ Gas Detectors ☐ UV Transmittance Analyzer ☐ None ☐ Other *(please specify)*_______
Automation and Control *(Backup / Secondary or Multipurpose Units)*	**Automation and Control *(Backup / Secondary or Multipurpose Units)***	**Automation and Control *(Backup / Secondary or Multipurpose Units)***	**Automation and Control *(Backup / Secondary or Multipurpose Units)***
☐ Dual Power Source ☐ Dual PLC's ☐ Portable or In-Place Generators ☐ Multiple Control Strategies ☐ DCS/SCADA System including telemetry ☐ Dual Field Mounted Control Boxes ☐ None ☐ Other *(please specify)*_______	☐ Dual Power Source ☐ Dual PLC's ☐ Portable or In-Place Generators ☐ Multiple Control Strategies ☐ DCS/SCADA System including telemetry ☐ Dual Field Mounted Control Boxes ☐ None ☐ Other *(please specify)*_______	☐ Dual Power Source ☐ Dual PLC's ☐ Portable or In-Place Generators ☐ Multiple Control Strategies ☐ DCS/SCADA System including telemetry ☐ Dual Field Mounted Control Boxes ☐ None ☐ Other *(please specify)*_______	☐ Dual Power Source ☐ Dual PLC's ☐ Portable or In-Place Generators ☐ Multiple Control Strategies ☐ DCS/SCADA System including telemetry ☐ Dual Field Mounted Control Boxes ☐ None ☐ Other *(please specify)*_______

Please Mail for Fax: 702-568-1446 Survey Responses to ITA by November 30, 2001 or Answer This Survey Online at www.instrument.org

REDUNDANCY DESIGN PRACTICES BY TREATMENT PROCESS

Please Identify Redundancy Design Practices for Your Treatment Process Areas (check all boxes that apply)

Secondary **(attached growth, suspended growth, etc.)** *(please check all redundant design practices that apply)*	**Tertiary / Advanced Treatment** **(filtration, nutrient removal, etc.)** *(please check all redundant design practices that apply)*	**Solids Processing** **(digestion, thickening, dewatering, composting, etc.)** *(please check all redundant design practices that apply)*	**Disinfection and Effluent** **(chlorination, dechlorination, UV, ozone, etc.)** *(please check all redundant design practices that apply)*
Reasons for Redundancy (Federal, State/Provincial, or Regional Regulations, etc.)	**Reasons for Redundancy (Federal, State/Provincial, or Regional Regulations, etc.)**	**Reasons for Redundancy (Federal, State/Provincial, or Regional Regulations, etc.)**	**Reasons for Redundancy (Federal, State/Provincial, or Regional Regulations, etc.)**
☐ **Discharge Permit Regulations** *(i.e., NPDES – National Pollutant Discharge Elimination System Permit Requirements)* ☐ **Safety Regulations** *(Occupational Health and Safety Admin.–OSHA; National Fire Protection Assoc –NFPA; Uniform Fire Code –UFC; National Institute of Occupational Safety & Health –NIOSH, etc.)* ☐ **In-House Operations & Maintenance Requirements** ☐ **IEEE Standard 493** *(Institute of Electrical and Electronic Engineers recommended practice for reliable electrical system design)* ☐ **Emergency / Disaster Planning Regulations/ Requirements** ☐ **None** ☐ **Other** *(please specify the name of the regulation that applies)* ________________	☐ **Discharge Permit Regulations** *(i.e., NPDES – National Pollutant Discharge Elimination System Permit Requirements)* ☐ **Safety Regulations** *(Occupational Health and Safety Admin.–OSHA; National Fire Protection Assoc –NFPA; Uniform Fire Code –UFC; National Institute of Occupational Safety & Health –NIOSH, etc.)* ☐ **In-House Operations & Maintenance Requirements** ☐ **IEEE Standard 493** *(Institute of Electrical and Electronic Engineers recommended practice for reliable electrical system design)* ☐ **Emergency / Disaster Planning Regulations/ Requirements** ☐ **None** ☐ **Other** *(please specify the name of the regulation that applies)* ________________	☐ **Solids Processing Regulations** *(i.e., Federal – US EPA 503 Regs, State/Province, Regional)* ☐ **Discharge Permit Regulations** *(i.e., NPDES – National Pollutant Discharge Elimination System Permit Requirements)* ☐ **Safety Regulations** *(Occupational Health and Safety Admin. –OSHA; National Fire Protection Assoc. –NFPA; Uniform Fire Code –UFC; National Institute of Occupational Safety & Health –NIOSH, etc.)* ☐ **In-House Operations & Maintenance Requirements** ☐ **IEEE Standard 493** *(Institute of Electrical and Electronic Engineers recommended practice for reliable electrical system design)* ☐ **Emergency / Disaster Planning Regulations/ Requirements** ☐ **None** ☐ **Other** *(please specify the name of the regulation that applies)* ________________	☐ **Discharge Permit Regulations** *(i.e., NPDES – National Pollutant Discharge Elimination System Permit Requirements)* ☐ **Safety Regulations** *(Occupational Health and Safety Admin. –OSHA; National Fire Protection Assoc. –NFPA; Uniform Fire Code –UFC; National Institute of Occupational Safety & Health –NIOSH, etc.)* ☐ **In-House Operations & Maintenance Requirements** ☐ **IEEE Standard 493** *(Institute of Electrical and Electronic Engineers recommended practice for reliable electrical system design)* ☐ **Emergency / Disaster Planning Regulations/ Requirements** ☐ **None** ☐ **Other** *(please specify the name of the regulation that applies)* ________________

Please Mail for Fax: 702-568-1446 Survey Responses to ITA by November 30, 2001 or Answer This Survey Online at www.instrument.org

REDUNDANCY DESIGN PRACTICES BY TREATMENT PROCESS

Please Identify Redundancy Design Practices for Your Treatment Process Areas (check all boxes that apply)

Secondary (attached growth, suspended growth, etc.) *(please check all redundant design practices that apply)*	**Tertiary / Advanced Treatment (filtration, nutrient removal, etc.)** *(please check all redundant design practices that apply)*	**Solids Processing (digestion, thickening, dewatering, composting, etc.)** *(please check all redundant design practices that apply)*	**Disinfection and Effluent (chlorination, dechlorination, UV, ozone, etc.)** *(please check all redundant design practices that apply)*
Redundancy Ratings	**Redundancy Ratings**	**Redundancy Ratings**	**Redundancy Ratings**
For the Redundancy Design Practices selected above for this treatment process, please rate the effectiveness of implementing these practices: ☐ **Extremely Effective** – Consistently use redundant systems and have been able to reduce failures with these practices. ☐ **Somewhat Effective** – Occasionally use redundant systems for preventative maintenance not for system failures. ☐ **Currently Not Necessary** – The redundant systems of this treatment process area is not currently used but may be used with future growth requirements. ☐ **Never Used** – The redundant systems have never been used and do not foresee a future use.	***For the Redundancy Design Practices selected above for this treatment process, please rate the effectiveness of implementing these practices:*** ☐ **Extremely Effective** – Consistently use redundant systems and have been able to reduce failures with these practices. ☐ **Somewhat Effective** – Occasionally use redundant systems for preventative maintenance not for system failures. ☐ **Currently Not Necessary** – The redundant systems of this treatment process area is not currently used but may be used with future growth requirements. ☐ **Never Used** – The redundant systems have never been used and do not foresee a future use.	***For the Redundancy Design Practices selected above for this treatment process, please rate the effectiveness of implementing these practices:*** ☐ **Extremely Effective** – Consistently use redundant systems and have been able to reduce failures with these practices. ☐ **Somewhat Effective** – Occasionally use redundant systems for preventative maintenance not for system failures. ☐ **Currently Not Necessary** – The redundant systems of this treatment process area is not currently used but may be used with future growth requirements. ☐ **Never Used** – The redundant systems have never been used and do not foresee a future use.	***For the Redundancy Design Practices selected above for this treatment process, please rate the effectiveness of implementing these practices:*** ☐ **Extremely Effective** – Consistently use redundant systems and have been able to reduce failures with these practices. ☐ **Somewhat Effective** – Occasionally use redundant systems for preventative maintenance not for system failures. ☐ **Currently Not Necessary** – The redundant systems of this treatment process area is not currently used but may be used with future growth requirements. ☐ **Never Used** – The redundant systems have never been used and do not foresee a future use.
Comments	**Comments**	**Comments**	**Comments**

Please provide any general comments and/or identify alternative redundancy design practices. Attach additional sheets for comments if necessary.

Please Mail for Fax: 702-568-1446 Survey Responses to ITA by November 30, 2001 or Answer This Survey Online at www.instrument.org

APPENDIX B

SURVEYED TREATMENT FACILITIES

Reference survey data is provided in a tabular format in this appendix for the listing of surveyed treatment facilities, as provided in Table B-1.

Table B-1. Surveyed Treatment Facility's Geographical Information.

Organization Name	Treatment Facility Name	City	State/ Province	Country
Anchorage Water and Wastewater Utility	3 ww plants ... eagle river/pt. woronzof /girdwood	Anchorage	Alaska	USA
Central Contra Costa Sanitary District	Central Contra Costa Sanitary District	Martinez	California	USA
City & County of Honolulu	Honouliuli Wastewater Treatment Plant	Honolulu	Hawaii	USA
City of Detroit	Detroit Wastewater Treatment Plant	Detroit	Michigan	USA
City of Edmonton	Gold Bar Wastewater Treatment Plant	Edmonton	Alberta	Canada
City of Eugene	Eugene/Springfield Water Pollution Control Facility	Eugene	Oregon	USA
City of Guelph	City of Guelph Wastewater Treatment Plant	Guelph	Ontario	Canada
City of Indianapolis	Belmont AWT/Sewer Collections	Indianapolis	Indiana	USA
City of Montreal	Montreal Wastewater Treatment Plant	Montreal	Quebec	Canada
City of Owosso	Owosso Mid-County WWTP	Owosso	Michigan	USA
City of Phoenix	91st Avenue Wastewater Treatment Plant	Phoenix	Arizona	USA
City of Roseville	Dry Creek WWTP	Roseville	California	USA
City of Santa Barbara	El Estero WWTP	Santa Barbara	California	USA
City of Stuart	City of Stuart Wastewater Treatment Facility	Stuart	Florida	USA
City of Toronto	Ashbridges Bay Treatment Plant	Toronto	Ontario	Canada
City of Winnipeg	North Plant	Winnipeg	Manitoba	Canada
City of Yuba City	Yuba City Wastewater Treatment Facility	Yuba City	California	USA
Columbus Water Works	South Columbus Water Resource Facility	Columbus	Georgia	USA
County Sanitation Districts of Los Angeles County	Joint Water Pollution Control Plant	Whittier	California	USA
Dow Chemical	Environmental Operations	Port Lavala	Texas	USA
Fairfax County Government	Noman Cole, Jr. Pollution Control Plant	Lorton	Virginia	USA
Greater Vancouver Regional District	Annacis Island Wastewater Treatment Plant	Burnaby	British Columbia	Canada
Gulf Coast Waste Disposal Authority	Odessa South	Houston	Texas	USA
Gulf Coast Waste Disposal Authority	Washburn Tunnel Facility	Pasadena	Texas	USA

Table B-1. Surveyed Treatment Facility's Geographical Information (continued).

Organization Name	Treatment Facility Name	City	State/ Province	Country
King County Wastewater Treatment Division	West Point Treatment Plant	Seattle	Washington	USA
Little Rock Wastewater	Fourche Creek Plant	Little Rock	Arkansas	USA
Madison Metropolitan Sewerage District	Nine Springs Wastewater Treatment Plant	Madison	Wisconsin	USA
Metro Watewater Reclamation District	Central Treatment Plant	Denver	Colorado	USA
Metropolitan Council Environmental Services	Metropolitan Wastewater Treatment Plant	St. Paul	Minnesota	USA
Metropolitan Water Reclamation District of Greater Chicago	Calumet Water Reclamation Plant	Chicago	Illinois	USA
Metropolitan Water Reclamation District of Greater Chicago	James C. Kirie Water Reclamation Plant	Des Plaines	Illinois	USA
Metropolitan Water Reclamation District of Greater Chicago	John E. Egan WRP	Schaumburg	Illinois	USA
Metropolitan Water Reclamation District of Greater Chicago	North Side WRP	Skokie	Illinois	USA
Metropolitan Water Reclamation District of Greater Chicago	Stickney Water Reclamation Plant	Cicero	Illinois	USA
Metrrpolitan St. Louis Sewer District	5 WWTPs and 250 pump stations	St. Louis	Missouri	USA
Milkwaukee Metropolitain Sewerage District	Jones Island WWTP	Milwaukee	Wisconsin	USA
Milwaukee Metropolitain Sewerage District	South Shore WWTP	Milwaukee	Wisconsin	USA
MWRDGC	Hanover Park WRP	Hanover Park	Illinois	USA
Northeast Ohio Regional Sewer District	Southerly WWTP	Cleveland	Ohio	USA
Orange County Utilities	South Water Reclamation Facility	Orlando	Florida	USA
Reedy Creek Improvement District	Water Reclamation Facility	Lake Buena Vista	Florida	USA
Sacramento Regional County Sanitation District	Sacramento Regional Wastewater Treatment Plant	Elk Grove	California	USA
Sanitary District of Decatur	Sanitary District of Decatur	Decatur	Illinois	USA
The City of Calgary	Bonnybrook Wastewater Treatment Plant	Calgary	Alberta	Canada
Tulsa Metropolitan Utility Authority	Southside Wastewater Treatment Plant	Tulsa	Oklahoma	USA
Western Lake Superior Sanitary District	Main Treatment Plant	Duluth	Minnesota	USA

GENERAL TREATMENT PLANT PROFILE GRAPHS

The following graphs provide an overview of the general treatment plant profiles of the surveyed treatment facilities. Detailed information for these graphs is contained in Appendix D in a tabular format.

C.1 General Treatment Plant Profiles

The majority of the plants participating in this survey treat wastewater (89%) followed by water treatment (7%) and industrial treatment (4%) as shown in Figure C-1. In addition, Table C-1 in Appendix C notes that of the 89% that treat wastewater, two treat both water and wastewater.

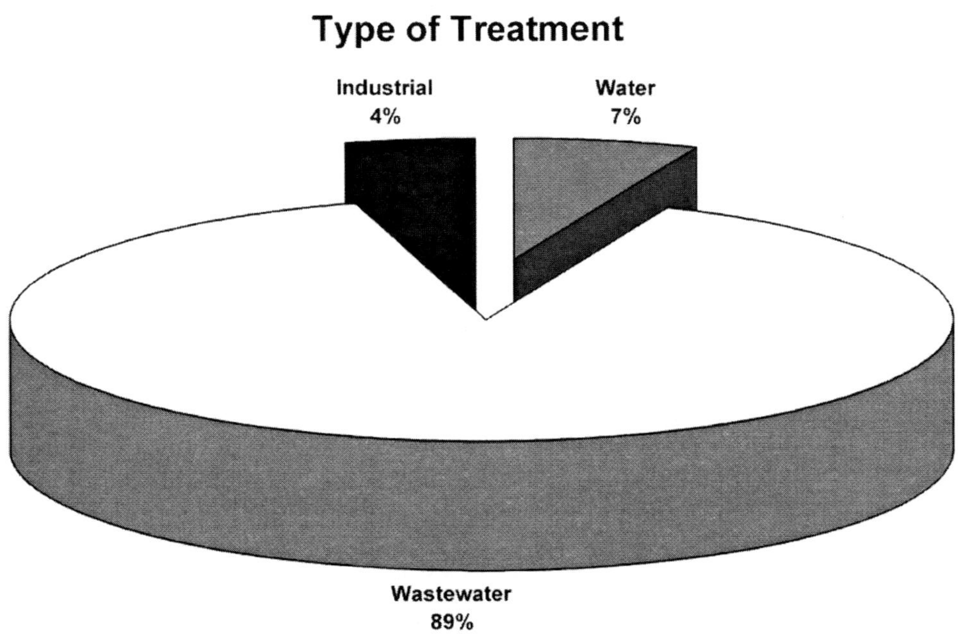

Figure C-1. Type of Treatment Plants.

Examining Figure C-2 reveals that 24% of the respondents reported using advanced wastewater treatment. This data is significant because one fourth of the respondents may have a higher level of redundancy for their advanced treatment process. Typically advanced wastewater treatment processes are more complicated and have numerous components to control.

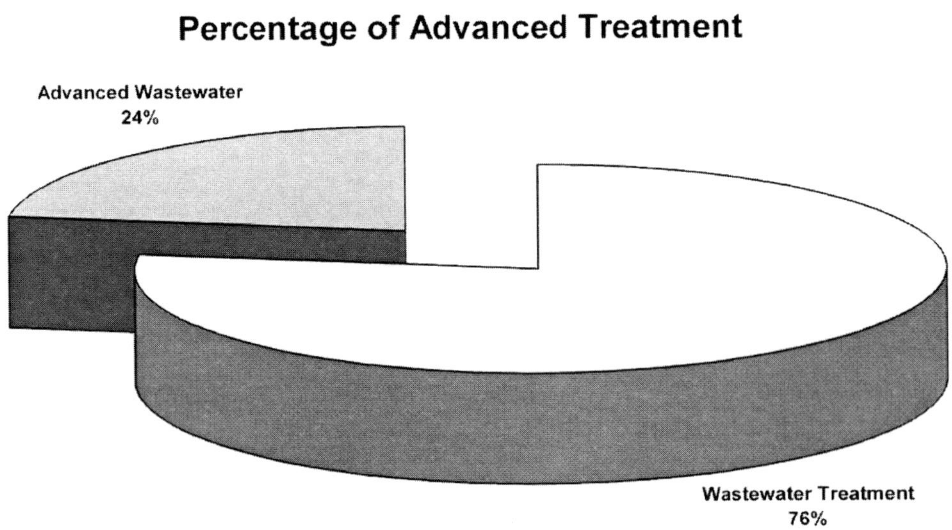

Figure C-2. Percentage of Advanced Treatment.

Figure C-3 shows that 98% of the survey respondents report that they have publicly owned facilities. Private facilities represent only 2% of the survey data. This large majority of publicly owned treatment facilities fairly represent the typical ownership of treatment facilities across the United States and Canada.

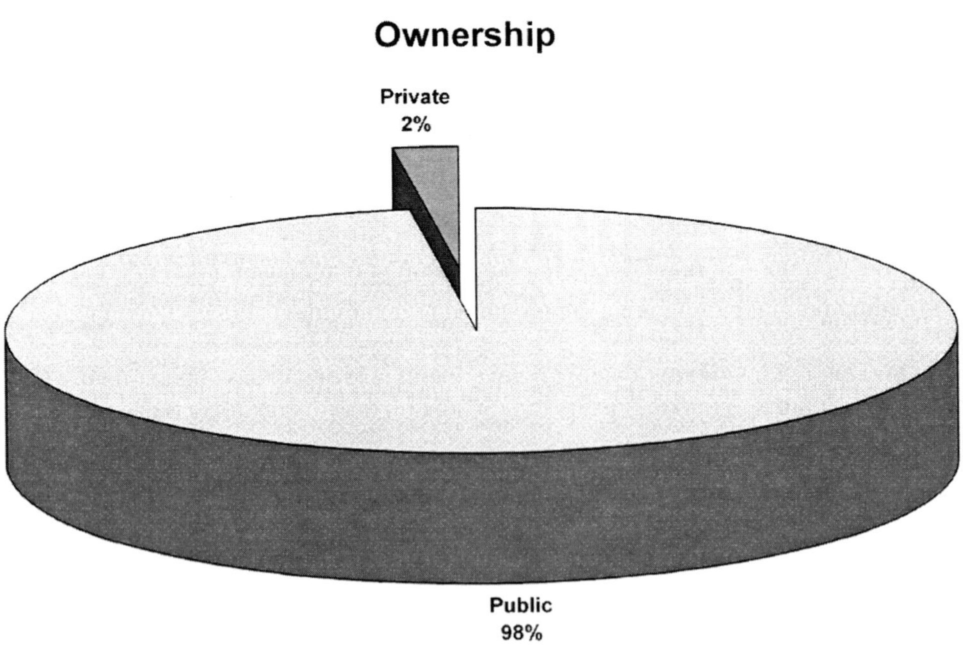

Figure C-3. Ownership.

Respondents to the survey report 92% of the facilities have public operational responsibilities as shown in Figure C-4. Contract operations are used by 6% of the treatment facilities that responded to the survey and private operational responsibility accounts for only 2% of the survey responses. Public operational responsibilities typifies the majority of operational and maintenance activities.

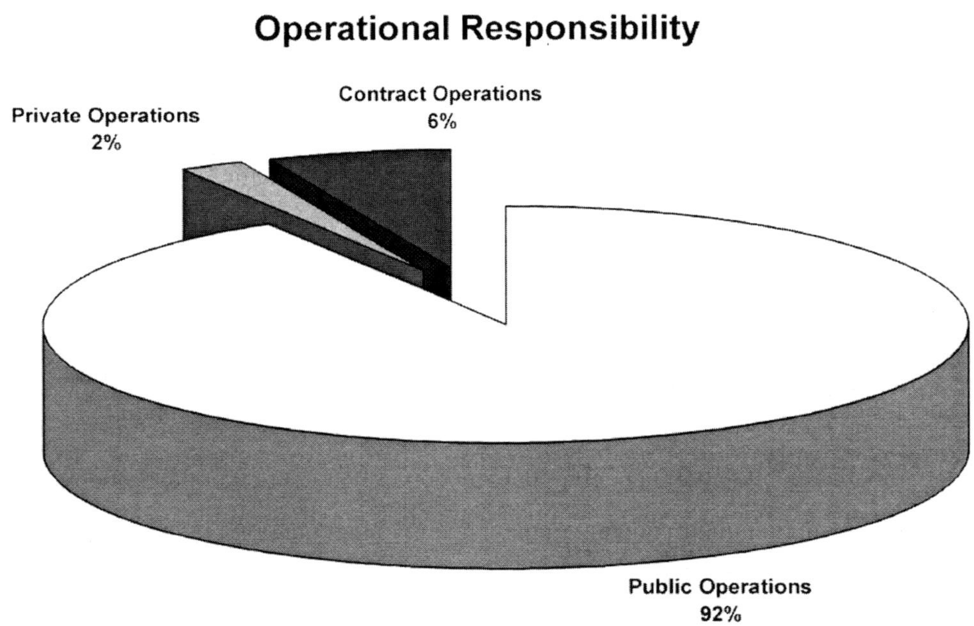

Figure C-4. Operational Responsibility.

Of the treatment facilities that responded to the survey, 98% reported attended operation for 7 days per week and only 2% of the facilities reported three to six days of attended operation as displayed in Figure C-5.

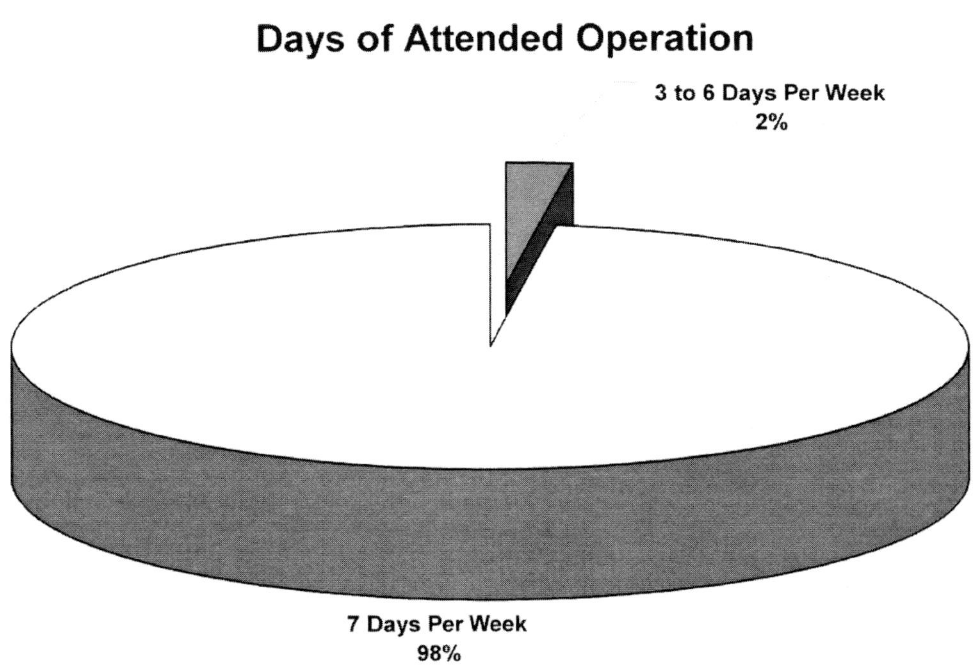

Figure C-5. Days of Attended Operation.

Figure C-6 shows 96% of the reporting facilities have 24 hours per day of attended operation. The remaining facilities that responded were evenly split at 2% with 10 and 16 hours of attended operation respectively.

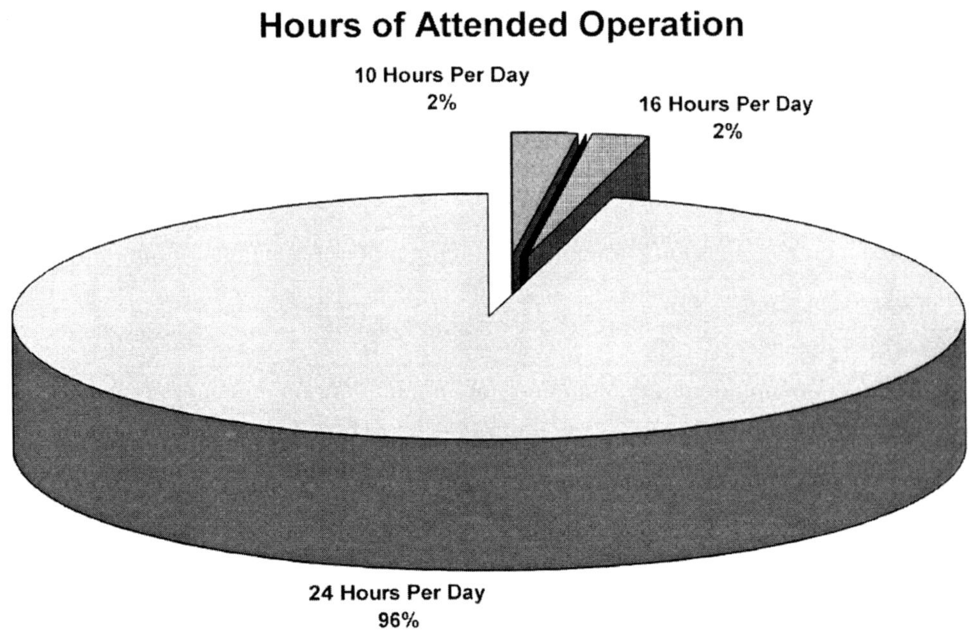

Figure C-6. Hours of Attended Operation.

Twenty-four of the total survey respondents reported sanitary sewer only as their influent flow source. Another 21 respondents reported having a combination of sanitary sewers and storm water for their influent flow source as shown in Figure C-7. Heavy storm flows combined with sanitary flows will influence how a utility handles collection system redundancy issues. Three facilities reported industrial influent flows.

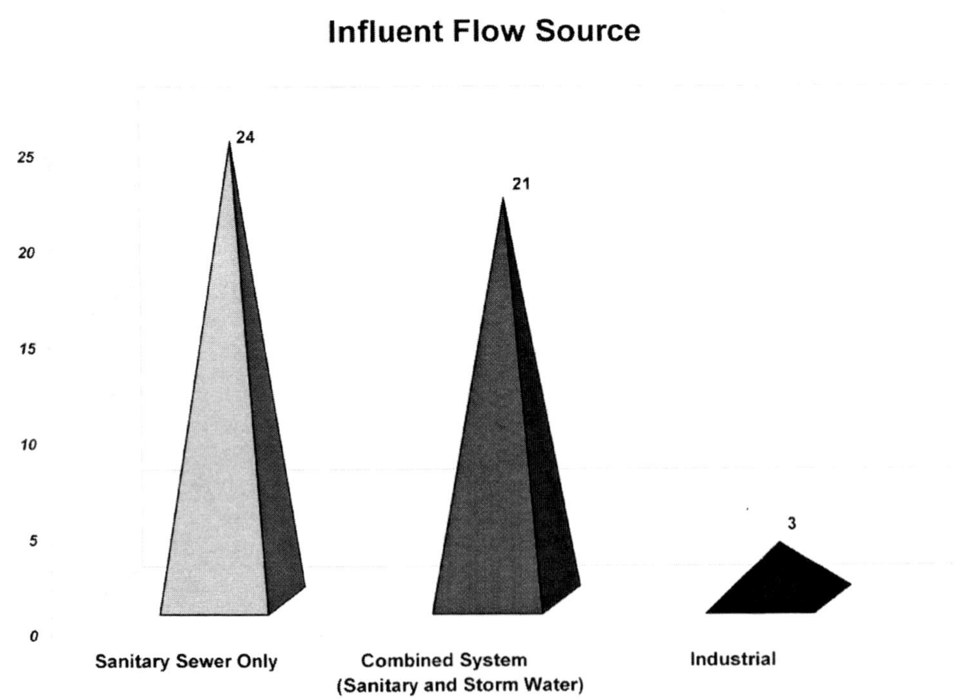

Figure C-7. Influent Flow Source.

Influent flow characteristics are shown in Figure C-8. Two facilities report primarily industrial flows and eight facilities have primarily municipal influent flows. 15 facilities report municipal and industrial flows, while 22 facilities report municipal flow combined with industrial pretreatment flow characteristics.

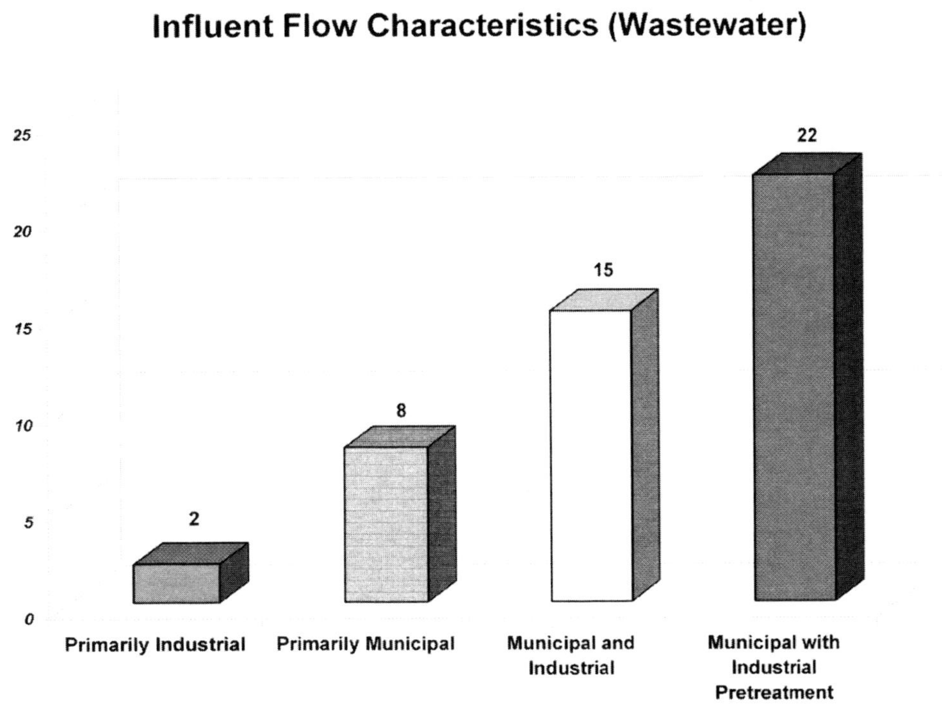

Figure C-8. Influent Flow Characteristics.

Treatment level is reported in Figure C-9. Secondary suspended growth accounts for 80% of the facilities that responded, whereas, secondary attached growth is reported at 10%. Approximately 6% of the survey respondents reported having only primary treatment and 4% of the facilities have other levels of treatment, such as trickling filter and anaerobic attached.

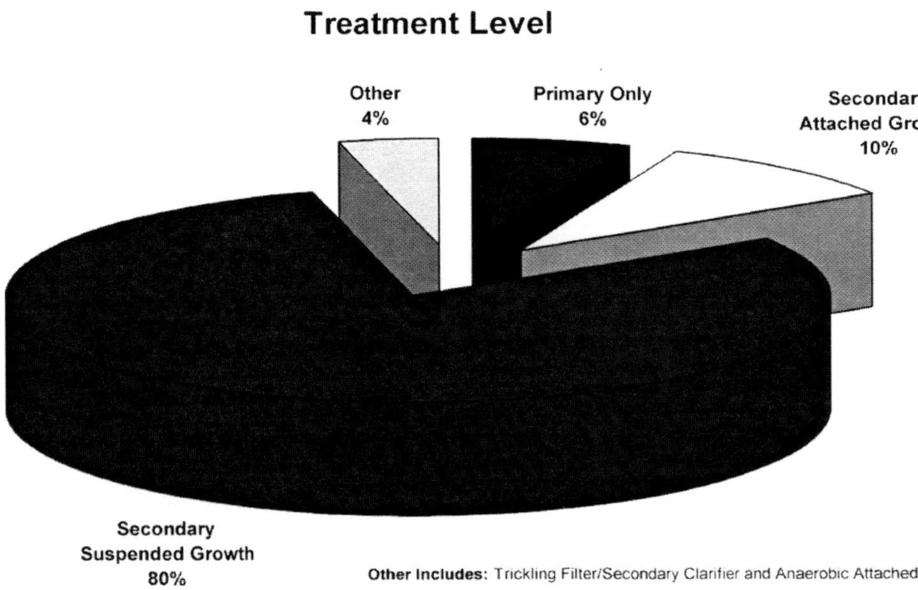

Figure C-9. Treatment Level.

The project survey requested treatment facilities to report how they provided disinfection for their effluent. The majority of facilities, 45%, reported they use chlorination and dechlorination treatment processes. Chlorination only is reported by 29% of the facilities, whereas 12% of the treatment facilities reported having no disinfection processes while 10% reported using ultraviolet light for disinfection. Other types of disinfection processes were reported by 4% of the facilities and include disinfection using sulfur dioxide and disinfection with sulfuric acid pretreatment.

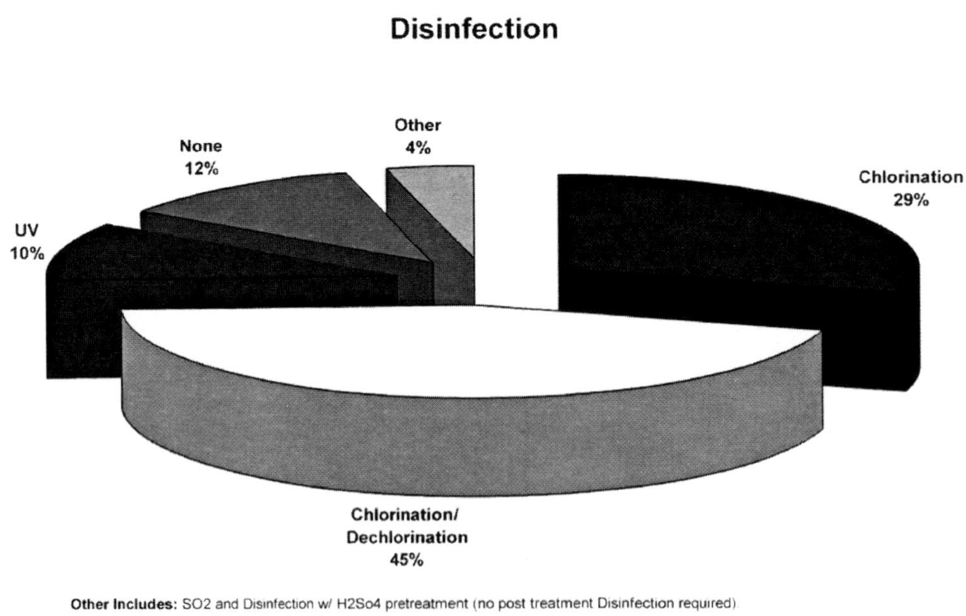

Figure C-10. Disinfection.

For the surveyed treatment facilities with advanced treatment, half (50%) reported using nutrient removal and the other half reported using filtration (44%) and wetland or no advanced treatment (6%).

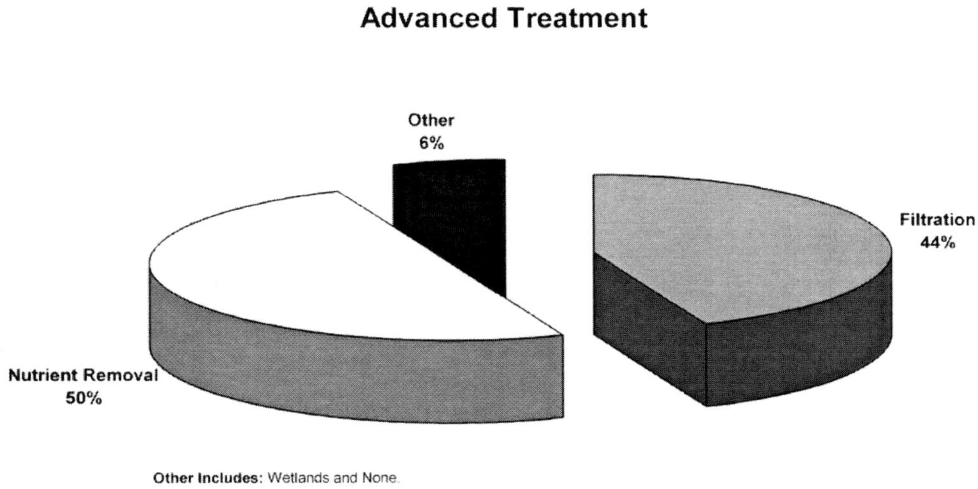

Figure C-11. Advanced Treatment.

Figure C-12 indicates that 83% of the facilities reported they are not part of a centralized monitoring system. However, 17% reported their facilities are part of a centralized monitoring system.

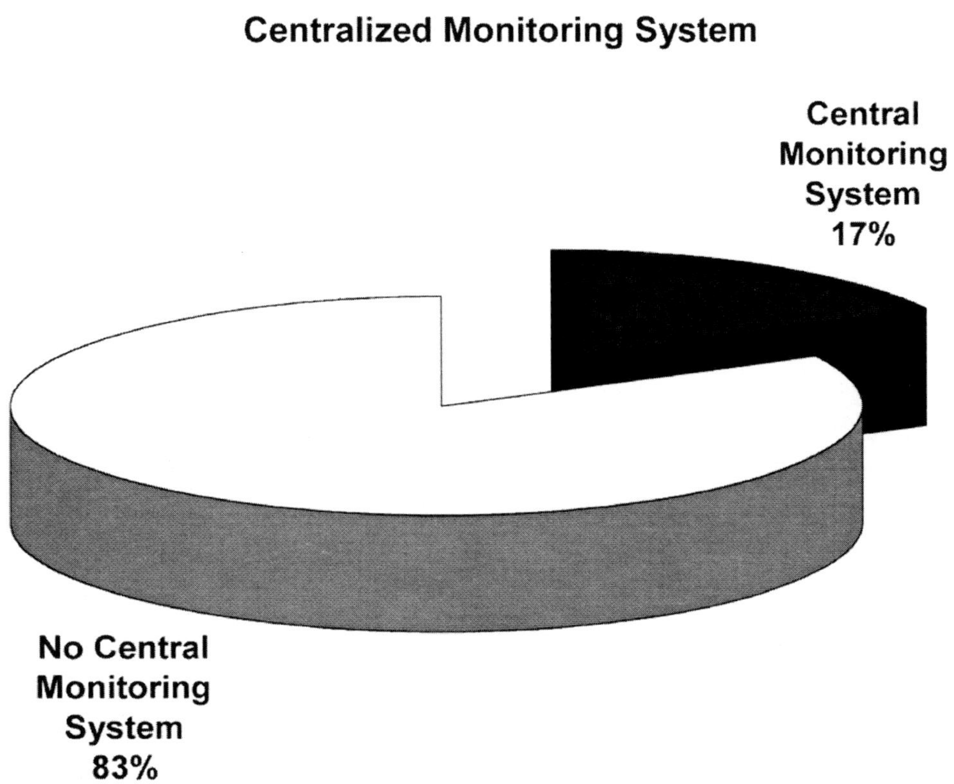

Figure C-12. Centralized Monitoring System.

One of the necessary tools to efficiently operate treatment plants is a control and automation system. Figure C-13 shows 89% of facilities reporting have some type of control and automation system, 11% reported their facility is not connected to an automation system.

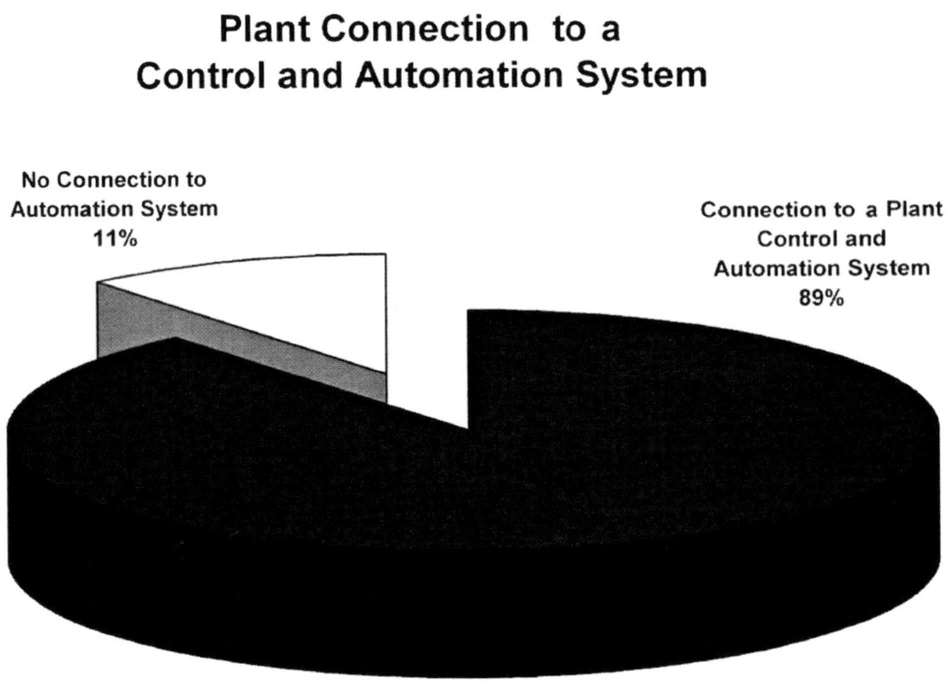

Figure C-13. Plant Connection to Control and Automation System.

APPENDIX D

General Treatment Facility Information Data Tables

Reference survey data is provided in a tabular format in this appendix for general treatment facility information. The following tables are included:

- Table D-1. General Plant Information
- Table D-2. Influent Flow
- Table D-3. Treatment Level
- Table D-4. Flow, Upgrade and Control Information
- Table D-5. Treatment Receiving Waters and Redundancy Design Practices
- Table D-6. General Redundancy Design Practices

Table D-1. General Plant Information.

ID	Type of Treatment Facility				Ownership		Operational Responsibility			Plant Days of Attended Operation			Plant Hours of Attended Operation						Staff
Plant ID No.	Water	Wastewater	Advanced Wastewater (Tertiary)	Industrial	Public	Private	Public	Private	Private (contract operations)	1-3 days per week	3-6 days per week	7 days per week	Less than 8 hours per day	8 hours per day	10 hours per day	12 hours per day	16 hours per day	24 hours per day	Total Operations and Maintenance Plant Staff
4		X	X		X		X					X						X	30
7		X			X		X					X						X	250
9		X			X		X					X						X	63
13		X			X		X					X						X	46
19	X		X				X					X						X	24
21		X			X		X					X					X		11
35		X			X		X					X						X	20
44	X				X		X				X	X			X			X	15
49		X			X		X					X						X	75
60		X	X		X		X					X						X	12
203		X			X		X					X						X	102
205		X			X		X					X						X	275 (157 maintenance; 86 operations; 32 elect/instru)
208		X	X		X		X					X						X	7
209		X			X		X					X						X	
210		X	X		X		X					X						X	18
211		X			X		X					X						X	50
212		X	X		X				X			X						X	
213		X			X		X					X						X	58
215		X			X		X					X						X	200
216			X	X		X		X				X						X	15
217		X			X		X					X						X	32 (25 staff plus 7

Table D-1. General Plant Information (continued).

ID	Type of Treatment Facility				Ownership		Operational Responsibility			Plant Days of Attended Operation			Plant Hours of Attended Operation						Staff
Plant ID No.	Water	Wastewater	Advanced Wastewater (Tertiary)	Industrial	Public	Private	Public	Private	Private (contract operations)	1-3 days per week	3-6 days per week	7 days per week	Less than 8 hours per day	8 hours per day	10 hours per day	12 hours per day	16 hours per day	24 hours per day	Total Operations and Maintenance Plant Staff
																			contract employees)
218		X			X		X					X						X	36
219		X	X		X		X					X						X	150
221		X			X		X					X						X	350
222		X			X		X					X						X	275
224		X			X		X					X						X	30
225		X	X		X		X					X						X	50
228		X			X		X					X						X	12
229		X	X		X		X					X						X	52
230		X			X		X					X						X	180
231		X			X		X					X						X	164
232		X			X		X					X						X	700
233		X			X		X					X						X	45
234		X			X				X			X						X	90
235		X			X				X			X						X	90
236		X			X		X					X						X	70
237	X		X		X		X					X						X	12
238		X			X		X					X						X	191
239		X			X		X					X						X	145
240			X	X	X		X					X						X	40
241		X			X		X					X						X	24 water & wastewater - 13 WW 11 Water

Table D-1. General Plant Information (continued).

ID	Type of Treatment Facility				Ownership		Operational Responsibility			Plant Days of Attended Operation			Plant Hours of Attended Operation						Staff
Plant ID No.	Water	Wastewater	Advanced Wastewater (Tertiary)	Industrial	Public	Private	Public	Private	Private (contract operations)	1-3 days per week	3-6 days per week	7 days per week	Less than 8 hours per day	8 hours per day	10 hours per day	12 hours per day	16 hours per day	24 hours per day	Total Operations and Maintenance Plant Staff
243		X	X		X		X					X						X	86
244		X			X		X					X						X	124
245		X			X		X					X						X	582
246		X			X		X					X						X	35
247		X	X		X		X					X						X	
Total	**3**	**41**	**14**	**2**	**44**	**1**	**42**	**1**	**3**	**0**	**1**	**46**	**0**	**0**	**1**	**0**	**1**	**45**	

Table D-2. Influent Flow.

ID	Influent Flow				Influent Flow Characteristics (wastewater)			
Plant ID No.	Sanitary Sewer Only	Combined System *(sanitary sewer and stormwater)*	Other	Other *(specify)*	Primarily Industrial	Primarily Municipal	Municipal and Industrial	Municipal with Industrial Pretreatment
4	X					X		
7	X	X						X
9		X					X	
13		X					X	
19	X					X		
21	X					X		
35	X							X
44	X					X		
49	X							X
60			X	60% Industrial			X	
203	X					X		
205	X						X	X
208		X						X
209	X						X	
210	X						X	
211		X					X	
212		X						X
213		X					X	
215		X					X	
216		X	X	Industrial	X			
217			X	Industrial Process WW by Force main	X			
218	X							X
219		X						X
221		X						X
222		X						X
224	X							X
225	X							X
228		X					X	
229		X						X
230		X						X
231		X						X
232		X						X
233	X							X
234	X						X	
235		X					X	
236	X							X
237	X					X		
238		X					X	
239	X					X		

Table D-2. Influent Flow (continued).

ID	Influent Flow				Influent Flow Characteristics (wastewater)			
Plant ID No.	Sanitary Sewer Only	Combined System *(sanitary sewer and stormwater)*	Other	Other *(specify)*	Primarily Industrial	Primarily Municipal	Municipal and Industrial	Municipal with Industrial Pretreatment
240	X						X	
241	X						X	
243	X					X		
244		X						X
245		X		Deep tunnel (TARP)				X
246	X							X
247	X							X
Total	**24**	**21**	**3**		**2**	**8**	**15**	**22**

Table D-3. Treatment Level.

ID	Treatment Level					Disinfection							Advanced Treatment			
Plant ID No.	Primary Treatment Only	Attached Growth	Suspended Growth	Other	Other Specify	Chlorination	Chlorination/ Dechlorination	UV	Ozone	None	Other	Other Specify	Filtration	Nutrient Removal	Other	Specify
4			X			X					X	SO2	X			
7		X				X	X									
9			X				X							X		
13								X						X		
19			X			X							X	X		
21			X							X				X		
35			X			X										
44	X		X			X	X						X			
49		X								X						
60			X				X						X	X		
203			X				X									
205			X			X										
208			X				X									
209			X					X						X		
210			X				X						X	X		
211			X					X								
212		X	X				X									
213		X	X	X	TF/SC		X									
215			X			X										
216				X	Anaerobic Attached					X					X	Wetlands
217			X								X	Disinfection w/ H2So4 pretreatment (no post treatment Disinfection required		X		
218			X			X							X			
219			X			X							X			
221			X				X							X		

Table D-3. Treatment Level (continued).

ID	Treatment Level					Disinfection							Advanced Treatment			
Plant ID No.	**Primary Treatment Only**	**Attached Growth**	**Suspended Growth**	**Other**	**Other Specify**	**Chlorination**	**Chlorination/ Dechlorination**	**UV**	**Ozone**	**None**	**Other**	**Other Specify**	**Filtration**	**Nutrient Removal**	**Other**	**Specify**
222			X							X						
224			X				X						X			
225			X				X						X			
228			X				X									
229			X				X						X			
230			X							X						
231	X													X		
232			X			X									X	None
233			X					X						X		
234			X				X							X		
235			X				X							X		
236			X				X									
237		X				X							X	X		
238	X		X			X	X									
239			X				X						X	X		
240			X			X	X						X			
241			X			X										
243			X					X					X			
244			X			X	X									
245			X							X						
246			X				X									
247			X				X							X		
Total	**3**	**5**	**40**	**2**		**15**	**23**	**5**	**0**	**6**	**2**		**14**	**16**	**2**	

Table D-4. Flow, Upgrade and Control Information.

ID	Flow			Year Began Service	Year Upgrade	Method of Design and Construction for Upgrade							Facility Instrumentation Connected to a Plant Control and Automation System		Treatment Facility Part of a Centralized Monitoring System for a Number of Treatment Plants	
Plant ID No.	Average Daily Flow (mgd)	Wet Weather Flow (mgd)	Maximum Capacity Design Flow (mgd)	Year Treatment Facility Established	Year of Most Recent Upgrade	In-House Design	Outside (Consultant) Design	In-House and Consultant Design	Design-Bid-Build with Single-Source	Construction Contractor Separate from Consultant	Other	Other Specify	Yes	No	Yes	No
4	14	24	45	1976	1992			X					X			X
7	300	700	650	1965	1996		X						X		X	
9	31	90	125	1921	1987		X			X			X			X
13	74	264	240	1956	2001			X		X			X			X
19	11	11	15	1970	2001			X					X			X
21	2	4	4	1964	1994		X		X	X			X			X
35	13	38	38	1983	1989		X						X		X	
44	30	30	60	1965	1995		X			X			X		X	
49	26	51	38	1982	1996		X							X		X
60	4	4	7	1997	2000			X					X			X
203	165	193	227	1966	1991		X			X			X			X
205	325	350	385	1928	2002	X							X			X
208	8	9	19	1978	1989		X			X			X			X
209	92	185	220	1932	1994		X			X			X			X
210	15	17	17	1927	2001		X			X			X			X
211	59	59	276	1935	1992			X					X		X	
212	125	270	300	1925	2001					X			X			X
213	116	132	288	1775	2000		X			X			X		X	
215	200	450	670	1943	2001			X	X				X			X
216	4	4	5	1994	1998		X							X		X

Table D-4. Flow, Upgrade and Control Information (continued).

ID	Flow			Year Began Service	Year Upgrade	Method of Design and Construction for Upgrade							Facility Instrumentation Connected to a Plant Control and Automation System		Treatment Facility Part of a Centralized Monitoring System for a Number of Treatment Plants	
Plant ID No.	Average Daily Flow (mgd)	Wet Weather Flow (mgd)	Maximum Capacity Design Flow (mgd)	Year Treatment Facility Established	Year of Most Recent Upgrade	In-House Design	Outside (Consultant) Design	In-House and Consultant Design	Design-Bid-Build with Single-Source	Construction Contractor Separate from Consultant	Other	Other Specify	Yes	No	Yes	No
217	20	28	60	1973	1994					X			X			X
218	26	27	31	1957	2001				X				X			X
219	107	710	735	1927	2002					X			X			X
221	220	700	735	1938	2002		X			X			X			X
222	250	430	430	1922	2002			X					X			X
224	8	22	22	1963	1998			X						X		X
225	26	125	150	1975	2002			X					X			X
228	32	45	83	1964	2001		X			X			X			X
229	32	72	110	1980	2002			X					X			X
230	265	450	450	1928	2002			X					X			X
231	682	1618	2016	1988	1988		X			X			X			X
232	750	1500	930	1941	2001		X			X				X		X
233	42	55	57	1928	1997			X					X			X
234	106	250	250	1968	2000		X			X			X			X
235	110	300	300	1925	2000		X			X			X			X
236	49	70	175	1984	2000		X			X			X			X
237	4	12	18	1983	1987			X		X				X		X
238	167	220	400	1982	2000			X					X		X	
239	45	65	54	1970	1995			X		X			X			X
240	38	46	160	1978	2001		X		X	X			X		X	
241	7	14	14	1971	2001		X						X			X

Table D-4. Flow, Upgrade and Control Information (continued).

ID	Flow			Year Began Service	Year Upgrade	Method of Design and Construction for Upgrade							Facility Instrumentation Connected to a Plant Control and Automation System		Treatment Facility Part of a Centralized Monitoring System for a Number of Treatment Plants	
Plant ID No.	Average Daily Flow (mgd)	Wet Weather Flow (mgd)	Maximum Capacity Design Flow (mgd)	Year Treatment Facility Established	Year of Most Recent Upgrade	In-House Design	Outside (Consultant) Design	In-House and Consultant Design	Design-Bid-Build with Single-Source	Construction Contractor Separate from Consultant	Other	Other Specify	Yes	No	Yes	No
243	45	55	53	1950	2000			X					X			X
244	126	131	300	1966	1995		X			X			X			X
245	800	1440	1440	1930	1973			X					X			X
246	32	84	84	1954	2002		X						X		X	
247	170	358	358	1960	2001		X						X			X
Total						1	24	17	4	23	0		41	5	8	38

Table D-5. Treatment Receiving Waters and Redundancy Design Practices.

ID	Receiving Waters Facilities Discharge Into									During Peak Flow Redundant Units Are			How Redundancy Design Practices Documented				
Plant ID No.	Drinking Water Receiving Stream	Recreation Receiving Stream	General Use Receiving Stream (not including drinking water or recreation use)	Percolation Ponds	Reclamation	Ocean	Lake	Other	Other Specify	Used	Not Used	Not Applicable – Redundant Units are Not in Place	Design Manual	Design Report	O & M Manual	Not Documented	Other Specify
4			X									X		X			
7	X	X	X							X				X			
9			X								X			X			
13	X	X								X	X			X			
19								X	Rapid Infiltration Basins and Reuse	X			X	X			
21								X	Deep Injection Well	X				X			
35		X									X			X			
44		X				X				X				X			
49					X	X				X			X	X			
60								X	Intermittant Stream	X	X				X		
203			X							X			X	X			
205						X					X		X	X			
208						X		X	Landscape			X		X		X	Policy and procedures
209		X								X			X	X			
210	X											X			X		
211		X								X	X				X	X	Historical Practices
212			X							X				X		X	Standard operating procedures
213								X	River Close to Ocean		X			X			
215							X			X			X				
216			X							X					X		

Water Environment Research Foundation

Table D-5. Treatment Receiving Waters and Redundancy Design Practices (continued).

ID	Receiving Waters Facilities Discharge Into									During Peak Flow Redundant Units Are			How Redundancy Design Practices Documented				
Plant ID No.	Drinking Water Receiving Stream	Recreation Receiving Stream	General Use Receiving Stream (not including drinking water or recreation use)	Percolation Ponds	Reclamation	Ocean	Lake	Other	Other Specify	Used	Not Used	Not Applicable – Redundant Units are Not in Place	Design Manual	Design Report	O & M Manual	Not Documented	Other Specify
217								X	Navigational water way		X				X		
218				X	X							X					
219			X							X			X	X			
221			X							X				X			
222								X	secondary contact	X			X	X			
224		X								X			X	X			
225			X							X			X	X			
228		X									X				X		
229			X							X			X	X			
230								X	secondary contact				X	X			
231		X									X				X		
232	X	X									X		X	X			
233		X								X			X				
234							X			X			X	X			
235							X			X			X	X			
236	X									X				X			
237		X								X			X	X			
238	X									X			X	X			
239								X	Creek to recreation waters	X			X	X			
240		X								X			X	X		X	Reliabilty standard - two times capacity plus a standby unit
241	X							X	Dillution > 250 :1	X					X		

Table D-5. Treatment Receiving Waters and Redundancy Design Practices (continued).

ID	Receiving Waters Facilities Discharge Into									During Peak Flow Redundant Units Are			How Redundancy Design Practices Documented				
Plant ID No.	Drinking Water Receiving Stream	Recreation Receiving Stream	General Use Receiving Stream (not including drinking water or recreation use)	Percolation Ponds	Reclamation	Ocean	Lake	Other	Other Specify	Used	Not Used	Not Applicable – Redundant Units are Not in Place	Design Manual	Design Report	O & M Manual	Not Documented	Other Specify
243								X	Estuary	X			X	X			
244								X	Salt Water - Puget Sound								
245								X	Secondary Contact	X				X			
246		X	X						Seasonal	X	X		X	X			
247			X		X						X		X	X			
Total	7	14	12	1	3	4	3	13		31	13	4	23	34	8	4	

Table D-6. General Redundancy Design Practices.

ID	Design Redundancy Practices In Place			Primary Driving Force for the Origin of Redundancy Design Practices										Other Means Treatment Facility Addresses Redundancy					
Plant ID No.	Yes	No	Not Aware	Federal Regulations	State/Province Regulations	Regional Regulations	Operational Efficiency	Maintenance Issues	Consultant Recommendations/ Design Guidelines	In-House or Organizational Specifications/ Policy/Design Guidelines	Emergency/ Disaster Planning	Other	Specify	Design Engineer Recommendations	Project Design Team Recommendations	Contractor Recommendations	Supplier Recommendations	Other	Specify
4	X						X	X						X					
7	X										X	X			X				
9	X						X	X	X			X			X				
13	X						X	X			X				X			X	OPS Recommendations
19	X			X	X							X							
21	X			X	X									X	X				
35	X						X								X				
44	X								X		X			X	X				
49							X	X				X		X					
60	X						X	X						X	X				
203	X			X	X		X	X	X					X	X				
205	X			X	X		X	X			X			X	X				
208	X						X					X		X				X	SD 709
209									X										
210																		X	No design redundancy program in place
211	X						X	X				X			X		X		
212	X			X	X		X	X						X	X				
213	X						X	X	X					X	X				
215	X										X				X				
216	X								X		X				X				
217	X						X	X				X							
218	X											X			X				
219	X			X	X	X	X	X	X					X	X				
221	X					X									X				

Table D-6. General Redundancy Design Practices (continued).

ID	Design Redundancy Practices In Place			Primary Driving Force for the Origin of Redundancy Design Practices										Other Means Treatment Facility Addresses Redundancy					
Plant ID No.	Yes	No	Not Aware	Federal Regulations	State/Province Regulations	Regional Regulations	Operational Efficiency	Maintenance Issues	Consultant Recommendations/ Design Guidelines	In-House or Organizational Specifications/ Policy/Design Guidelines	Emergency/ Disaster Planning	Other	Specify	Design Engineer Recommendations	Project Design Team Recommendations	Contractor Recommendations	Supplier Recommendations	Other	Specify
222	X				X		X	X				X			X				
224	X				X		X	X				X			X				
225	X				X		X	X				X			X				
228	X			X							X			X	X				
229	X				X		X	X				X			X				
230	X				X										X				
231	X							X						X					
232	X			X	X		X	X			X				X				
233	X				X		X	X	X		X				X				
234	X					X	X	X	X					X	X				
235	X					X	X	X	X					X	X				
236	X			X	X		X	X				X		X	X				
237	X			X	X	X													
238	X			X			X	X		X	X				X				
239	X			X	X	X	X	X	X			X		X	X				
240	X			X		X	X	X			X	X		X	X				
241	X							X	X		X			X					
243	X			X	X						X	X			X		X		
244	X				X									X	X				
245	X				X		X	X			X				X				
246	X			X	X				X		X			X	X				
247	X						X	X			X	X			X				
Total	43	0	0	15	20	7	28	28	13	1	16	17		21	36	0	2	3	

Collection Systems, Lift Stations, and Influent Flow

Reference survey data is provided in a tabular format in this appendix for treatment facility collection systems, lift stations, and influent flow. The following tables are included:

- Table E-1. Infrastructure and Equipment
- Table E-2. Instrumentation, Automation and Control
- Table E-3. Reasons for Redundancy, Redundancy Ratings, and General Comments

Key Code:

Safety Regulations	*Occupational Health and Safety Administration—OSHA; National Fire Protection Association—NFPA; Uniform Fire Code—UFC; National Institute of Occupational Safety & Health—NIOSH*
IEEE Standard 493	Institute of Electrical and Electronic Engineers recommended practice for reliable electrical system design
Extremely Effective	Consistently use redundant systems and have been able to reduce failures with these practices
Somewhat Effective	Occasionally use redundant systems for preventative maintenance not for system failures
Currently Not Necessary	The redundant systems of this treatment process area is not currently used but may be used with future growth requirements
Never Used	The redundant systems have never been used and do not foresee a future use

Table E-1. Infrastructure and Equipment Collection Systems, Lift Stations, and Influent Flow.

ID	Infrastructure													Equipment (backup or secondary)				
	Bypass Units						Backup/Secondary or Multipurpose Units											
Plant ID No.	Bypass Influent Line or Channel into Plant	Bypass lines at lift station(s)	Bypass Discharge to Local Water Body	None	Other	Specify	Influent Storage Basin/Tank	Collection System Storage Basin/Tank	Lift Station Storage Basin/Tank	Stormwater Storage Basin/Tank	None	Other	Specify	Pumps at Lift Stations	Portable Pumping	None	Other	Specify
4				X			X							X	X			
7	X						X		X					X				
9	X							X						X	X			
13	X									X				X				
19				X			X							X	X			Portable Generator
21														X	X			
35	X										X				X			
44	X	X					X							X	X			
49				X							X			X				
60				X			X			X				X				
203			X								X			X			X	Dual or triple electric power
205	X	X							X					X	X			
208				X							X			X	X			
209			X								X							
210	X										X			X	X			
211		X	X									X	Storage in sewerage (inline)	X	X		X	minimum of 2 pumps to meet required capacity
212					X	Siphon Barrel						X	Inflatable weir sluice gate	X	X			
213																		
215				X							X			X				
216		X									X			X	X			

Water Environment Research Foundation

Table E-1. Infrastructure and Equipment Collection Systems, Lift Stations, and Influent Flow (continued).

ID	Infrastructure													Equipment (backup or secondary)				
	Bypass Units						Backup/Secondary or Multipurpose Units											
Plant ID No.	Bypass Influent Line or Channel into Plant	Bypass lines at lift station(s)	Bypass Discharge to Local Water Body	None	Other	Specify	Influent Storage Basin/Tank	Collection System Storage Basin/Tank	Lift Station Storage Basin/Tank	Stormwater Storage Basin/Tank	None	Other	Specify	Pumps at Lift Stations	Portable Pumping	None	Other	Specify
217			X								X			X				
218				X							X			X	X			
219			X									X	deep tunnel project underway	X				
221				X							X			X				
222		X	X									X	tunnel and reservoir facility (TARP)	X				
224	X							X						X				
225			X								X			X				
228				X							X				X			
229					X	N/A	X							X				
230		X	X								X		tunnel and reservoir (TARP)	X				
231			X								X			X				
232			X					X						X				
233					X	We have some ability to route flows from an overloaded pump station to a pump station that is not over-loaded.					X			X	X			
234	X										X			X	X			
235	X										X			X	X			

Table E-1. Infrastructure and Equipment Collection Systems, Lift Stations, and Influent Flow (continued).

ID	Infrastructure													Equipment (backup or secondary)				
	Bypass Units						Backup/Secondary or Multipurpose Units											
Plant ID No.	Bypass Influent Line or Channel into Plant	Bypass lines at lift station(s)	Bypass Discharge to Local Water Body	None	Other	Specify	Influent Storage Basin/Tank	Collection System Storage Basin/Tank	Lift Station Storage Basin/Tank	Stormwater Storage Basin/Tank	None	Other	Specify	Pumps at Lift Stations	Portable Pumping	None	Other	Specify
236	X	X	X								X			X	X			
237		X									X			X	X			
238			X				X	X						X				
239				X			X							X	X			
240				X							X			X	X		X	Agreement with City to mutually take additional flows
241		X												X	X		X	Portable Not installed mobile pump station
243							X							X	X			
244		X	X					X	X	X				X	X			
245		X	X		X	bypass into TARP system		X					TARP Tunnel and Reservoir Plan)	X				
246				X					X					X	X			
247				X			X							X				
Total	**11**	**11**	**14**	**13**	**4**		**10**	**6**	**4**	**3**	**21**	**4**		**42**	**25**	**0**	**4**	

Table E-2. Instrumentation, Automation and Control Collection Systems, Lift Stations, and Influent Flow.

ID	Instrumentation (Backup/Secondary or Multipurpose Units)									Automation and Control (Backup/Secondary or Multipurpose Units)								
Plant ID No.	**Influent Flowmeters**	**Collection System Flowmeters**	**Wet Well Level Sensor**	**Wet Well Gas Detectors**	**Stormwater Flowmeters**	**Accelerometers or Displacement Monitors for Vibration Analysis**	**None**	**Other**	**Specify**	**Dual Power Source**	**Dual PLCs**	**Portable or In-Place Generators**	**Multiple Control Strategies**	**DCS/SCADA Systems including telemetry**	**Dual Field-Mounted Control Boxes**	**None**	**Other**	**Specify**
4							X					X		X				
7			X							X				X				
9			X							X		X	X	X				
13																		
19			X							X		X	X	X	X			
21	X		X									X		X				
35	X	X	X							X		X		X				
44			X	X						X		X		X				
49		X	X	X								X		X				
60							X			X							X	UPS
203			X							X		X	X	X				
205		X	X			X		X	most collection system flowmeters are portable	X		X	X	X				
208			X	X										X				
209							X					X				X		
210							X					X		X				
211							X	X	We have a number of portable flowmeters.								X	Alarm RTU (remote terminal unit) via phone line
212			X							X		X	X					
213																		
215	X											X				X		

Table E-2. Instrumentation, Automation and Control Collection Systems, Lift Stations, and Influent Flow (continued).

ID	Instrumentation (Backup/Secondary or Multipurpose Units)									Automation and Control (Backup/Secondary or Multipurpose Units)								
Plant ID No.	Influent Flowmeters	Collection System Flowmeters	Wet Well Level Sensor	Wet Well Gas Detectors	Stormwater Flowmeters	Accelerometers or Displacement Monitors for Vibration Analysis	None	Other	Specify	Dual Power Source	Dual PLCs	Portable or In-Place Generators	Multiple Control Strategies	DCS/SCADA Systems including telemetry	Dual Field-Mounted Control Boxes	None	Other	Specify
216							X									X		
217							X			X	X							
218			X	X								X		X				
219			X	X						X		X	X		X			
221							X					X						
222			X							X	X	X	X	X				
224							X			X			X					
225							X			X			X	X				
228			X							X			X					
229							X			X		X	X					
230			X							X		X	X	X				
231										X		X	X					
232			X							X	X							
233							X			X		X	X					
234							X			X		X					X	10-States Standard
235							X			X		X					X	10-States Standard
236			X							X	X	X	X	X				
237							X			X								
238										X				X				
239	X		X	X		X				X	X	X	X	X	X			
240			X							X		X	X	X				
241							X					X						

Table E-2. Instrumentation, Automation and Control Collection Systems, Lift Stations, and Influent Flow (continued).

ID	Instrumentation (Backup/Secondary or Multipurpose Units)									Automation and Control (Backup/Secondary or Multipurpose Units)								
Plant ID No.	Influent Flowmeters	Collection System Flowmeters	Wet Well Level Sensor	Wet Well Gas Detectors	Stormwater Flowmeters	Accelerometers or Displacement Monitors for Vibration Analysis	None	Other	Specify	Dual Power Source	Dual PLCs	Portable or In-Place Generators	Multiple Control Strategies	DCS/SCADA Systems including telemetry	Dual Field-Mounted Control Boxes	None	Other	Specify
243	X	X	X	X		X				X	X	X		X				
244			X							X		X		X	X			
245										X	X	X	X	X				
246		X	X	X						X		X	X	X				
247			X			X				X	X	X		X				
Total	5	5	24	8	0	4	16	2		32	8	32	19	25	4	3	4	

Table E-3. Reasons for Redundancy, Redundancy Ratings, and General Comments Collection Systems, Lift Stations, and Influent Flow.

ID	Reasons for Redundancy (*Federal, State/Provincial, or Regional Regulations*)											Redundancy Ratings				Comments
Plant ID No.	Flood Control District Regulations	Inflow/Infiltration (I/I) Regulations	Stormwater Regulations	Sanitary Sewer Overflow (SSO) Regulations	Safety Regulations	In-House Operations and Maintenance Requirements	IEEE Standard 493	Emergency/ Disaster Planning Regulations/ Requirements	None	Other	Specify	Extremely Effective	Somewhat Effective	Currently Not Necessary	Never Used	Comments
4						X						X				SCADA Monitoring and Backup Generators prevents sewage spills and upsets.
7						X		X				X				
9						X		X				X				
13																The above responses pertain to treatment facility only.
19					X			X					X			
21		X		X		X		X					X			
35						X							X			
44						X							X			
49				X		X		X				X				
60						X						X				
203		X		X	X	X						X				
205				X		X					most collection system flowmeters are portable	X				
208				X		X		X						X		
209			X	X	X							X				
210									X							No formal redundancy practices in place.
211						X					We have a number of portable flowmeters.		X			
212		X	X	X	X	X	X	X					X			A few extremely effective.

Table E-3. Reasons for Redundancy, Redundancy Ratings, and General Comments Collection Systems, Lift Stations, and Influent Flow (continued).

ID	Reasons for Redundancy (Federal, State/Provincial, or Regional Regulations)											Redundancy Ratings				Comments
Plant ID No.	Flood Control District Regulations	Inflow/Infiltration (I/I) Regulations	Stormwater Regulations	Sanitary Sewer Overflow (SSO) Regulations	Safety Regulations	In-House Operations and Maintenance Requirements	IEEE Standard 493	Emergency/ Disaster Planning Regulations/ Requirements	None	Other	Specify	Extremely Effective	Somewhat Effective	Currently Not Necessary	Never Used	Comments
213																Not applicable.
215						X				X		X				
216					X					X		X				
217						X		X				X				
218								X						X		
219					X		X									
221		X				X										
222					X	X		X					X			
224		X		X		X		X					X			
225				X		X		X					X			
228			X	X		X		X					X			
229					X	X		X					X			None
230						X		X					X			
231						X		X				X				
232						X							X			
233				X		X						X				
234										X		X				
235										X		X				
236				X		X		X				X				
237						X				X		X				
238																
239					X	X		X				X				None
240		X		X		X	X	X				X				

Table E-3—Reasons for Redundancy, Redundancy Ratings, and General Comments Collection Systems, Lift Stations, and Influent Flow (continued).

ID	Reasons for Redundancy (Federal, State/Provincial, or Regional Regulations)											Redundancy Ratings				Comments
Plant ID No.	Flood Control District Regulations	Inflow/Infiltration (I/I) Regulations	Stormwater Regulations	Sanitary Sewer Overflow (SSO) Regulations	Safety Regulations	In-House Operations and Maintenance Requirements	IEEE Standard 493	Emergency/ Disaster Planning Regulations/ Requirements	None	Other	Specify	Extremely Effective	Somewhat Effective	Currently Not Necessary	Never Used	Comments
241						X										Redundancy used only due to mechanical failure or extended power outage.
243	X	X	X	X	X	X	X	X					X			
244				X	X							X				
245	X		X			X		X				X				TARP (Tunnel and Reservoir Plan) is extremely effective for stormwater bypass and storage
246				X		X						X				
247						X				X		X				
Total	2	7	5	16	11	34	4	21	1	6		23	14	2	0	

APPENDIX F

INDUSTRIAL PRETREATMENT

Reference survey data is provided in a tabular format in this appendix for treatment facility industrial pretreatment processes. The following tables are included:

- Table F-1. Infrastructure and Equipment
- Table F-2. Instrumentation, Automation and Control
- Table F-3. Reasons for Redundancy, Redundancy Ratings, and General Comments

Key Code:

Safety Regulations	*Occupational Health and Safety Administration—OSHA; National Fire Protection Association—NFPA; Uniform Fire Code—UFC; National Institute of Occupational Safety & Health—NIOSH*
IEEE Standard 493	Institute of Electrical and Electronic Engineers recommended practice for reliable electrical system design
Extremely Effective	Consistently use redundant systems and have been able to reduce failures with these practices
Somewhat Effective	Occasionally use redundant systems for preventative maintenance not for system failures
Currently Not Necessary	The redundant systems of this treatment process area is not currently used but may be used with future growth requirements
Never Used	The redundant systems have never been used and do not foresee a future use

Table F-1. Infrastructure and Equipment Industrial Pretreatment.

ID	Infrastructure															Equipment									
	Bypass Units								Backup/Secondary or Multipurpose Units																
Plant ID No.	Flow Diversion Bypass Channel/Pipeline	Influent Flow Channel/ Pipeline	Gravity Separation or Sedimentation Bypass	Dissolved air Flotation (DAF)	Chemical Neutralization or Precipitation Bypass	None	Other	Specify	Flow Equalization Basin/Tank	Gravity Separation or Sedimentation Basin/Tank	Chemical Neutralization or Precipitation Basin/Tank	Chemical Feed	None	Other	Specify	Influent Pumps	Screens	Grinders	Chemical Feed Pumps	Sludge Pumps	Skimmers	Mixers	None	Other	Specify
4						X							X			X	X								
7						X							X			X			X	X					
9	X			X	X				X	X							X		X	X					
13																									
19																									
21						X							X												
35	X	X											X			X				X					
44																									
49						X			X							X	X	X	X	X	X	X			
60																									
203																									
205	X		X	X					X	X		X					X	X	X			X			
208						X							X			X		X		X					
209						X							X										X		
210						X							X			X	X			X					
211						X							X										X		
212						X							X										X		
213																									
215						X							X										X		
216			X										X			X			X						

Table F-1. Infrastructure and Equipment Industrial Pretreatment (continued).

ID	Infrastructure															Equipment									
	Bypass Units								Backup/Secondary or Multipurpose Units																
Plant ID No.	Flow Diversion Bypass Channel/Pipeline	Influent Flow Channel/ Pipeline	Gravity Separation or Sedimentation Bypass	Dissolved air Flotation (DAF)	Chemical Neutralization or Precipitation Bypass	None	Other	Specify	Flow Equalization Basin/Tank	Gravity Separation or Sedimentation Basin/Tank	Chemical Neutralization or Precipitation Basin/Tank	Chemical Feed	None	Other	Specify	Influent Pumps	Screens	Grinders	Chemical Feed Pumps	Sludge Pumps	Skimmers	Mixers	None	Other	Specify
217						X						X				X	X		X	X					
218																X	X		X	X					
219	X												X										X		
221																									
222																									
224																									
225																									
228																									
229																									
230																									
231																									
232							X	Not Applicable						X	Not Applicable										
233																									
234																									
235																								X	10-States Standard
236																									
237																									
238																X	X	X	X						
239	X	X							X							X	X	X	X	X	X	X			
240						X							X										X		

Table F-1. Infrastructure and Equipment Industrial Pretreatment (continued).

ID	Infrastructure															Equipment									
	Bypass Units								Backup/Secondary or Multipurpose Units																
Plant ID No.	Flow Diversion Bypass Channel/Pipeline	Influent Flow Channel/ Pipeline	Gravity Separation or Sedimentation Bypass	Dissolved air Flotation (DAF)	Chemical Neutralization or Precipitation Bypass	None	Other	Specify	Flow Equalization Basin/Tank	Gravity Separation or Sedimentation Basin/Tank	Chemical Neutralization or Precipitation Basin/Tank	Chemical Feed	None	Other	Specify	Influent Pumps	Screens	Grinders	Chemical Feed Pumps	Sludge Pumps	Skimmers	Mixers	None	Other	Specify
241						X							X										X		
243																									
244																									
245																									
246																									
247						X							X			X	X	X	X	X		X			
Total	5	2	2	2	1	14	1		4	2	0	2	15	1		12	10	6	10	10	2	4	7	1	

Table F-2. Instrumentation, Automation and Control Industrial Pretreatment.

ID	Instrumentation (Backup/Secondary or Multipurpose Units)										Automation and Control (Backup/Secondary or Multipurpose Units)								
Plant ID No.	pH Analyzers	Analytical Chemistry Analyzers *inorganic/organic compounds)*	Flowmeters	Level Sensors	Temperature Sensors	Oil and Grease Detectors	Coagulant Controller	None	Other	Specify	Dual Power Source	Dual PLCs	Portable or In-Place Generators	Multiple Control Strategies	DCS/SCADA System including telemetry	Dual Field-Mounted Control Boxes	None	Other	Specify
4								X									X		
7								X									X		
9								X			X		X	X	X				
13																			
19																			
21																			
35								X			X		X		X				
44																			
49	X	X	X	X									X		X				
60																			
203																			
205	X	X	X	X							X		X						
208								X							X				
209								X									X		
210								X											
211								X									X		
212								X									X		
213																			
215								X									X		
216								X									X		
217	X										X	X							

Table F-2. Instrumentation, Automation and Control Industrial Pretreatment (continued).

ID	Instrumentation (*Backup/Secondary or Multipurpose Units*)										Automation and Control (*Backup/Secondary or Multipurpose Units*)								
Plant ID No.	pH Analyzers	Analytical Chemistry Analyzers *inorganic/organic compounds)*	Flowmeters	Level Sensors	Temperature Sensors	Oil and Grease Detectors	Coagulant Controller	None	Other	Specify	Dual Power Source	Dual PLCs	Portable or In-Place Generators	Multiple Control Strategies	DCS/SCADA System including telemetry	Dual Field-Mounted Control Boxes	None	Other	Specify
218	X		X																
219								X									X		
221																			
222																			
224																			
225																			
228																			
229																			
230																			
231																			
232																			
233																			
234																			
235																			
236																			
237																			
238																			
239	X	X	X	X	X						X	X	X	X	X	X			
240								X									X		
241								X									X		
243																			
244																			

Table F-2. Instrumentation, Automation and Control Industrial Pretreatment (continued).

ID	Instrumentation (Backup/Secondary or Multipurpose Units)										Automation and Control (Backup/Secondary or Multipurpose Units)								
Plant ID No.	pH Analyzers	Analytical Chemistry Analyzers *inorganic/organic compounds)*	Flowmeters	Level Sensors	Temperature Sensors	Oil and Grease Detectors	Coagulant Controller	None	Other	Specify	Dual Power Source	Dual PLCs	Portable or In-Place Generators	Multiple Control Strategies	DCS/SCADA System including telemetry	Dual Field-Mounted Control Boxes	None	Other	Specify
245																			
246																			
247								X									X		
Total	5	3	4	3	1	0	0	15	0		5	2	5	2	5	1	11	0	

Table F-3. Reasons for Redundancy, Redundancy Ratings, and General Comments Industrial Pretreatment.

ID	Reasons for Redundancy (Federal, State/Provincial, or Regional Regulations)										Redundancy Ratings				Comments
Plant ID No.	General Pretreatment Regulations	In-House Pollution Prevention Program	Discharge Permit Regulations	Safety Regulations	In-House Operations & Maintenance Requirements	IEEE Standard 493	Emergency/Disaster Planning Regulations/ Requirements	None	Other	Specify	Extremely Effective	Somewhat Effective	Currently Not Necessary	Never Used	Comments
4								X							
7															
9					X		X				X				
13															This is responsibility of individual businesses and industries and varies considerably. Our staff inspect, sample and analyze discharges to municipal system to ensure compliance.
19															Our industrial users/dischargers do not practice any pretreatment, due to the nature of their discharges. We have an Industrial Waste Pretreatment Program because federal regulations require that all facilities that use reclaimed water have such a program.
21															
35					X						X				
44															
49			X		X		X				X				
60															
203															
205			X		X							X			Response relates to the redundant design practices encountered in the infrastructure of industrial wastewater dischargers within the districts' jurisdiction.
208			X		X		X							X	
209		X										X			
210								X							No formal redundancy practices in place.

Water Environment Research Foundation

Table F-3. Reasons for Redundancy, Redundancy Ratings, and General Comments Industrial Pretreatment (continued).

ID	Reasons for Redundancy (*Federal, State/Provincial, or Regional Regulations*)										Redundancy Ratings				Comments
Plant ID No.	General Pretreatment Regulations	In-House Pollution Prevention Program	Discharge Permit Regulations	Safety Regulations	In-House Operations & Maintenance Requirements	IEEE Standard 493	Emergency/Disaster Planning Regulations/ Requirements	None	Other	Specify	Extremely Effective	Somewhat Effective	Currently Not Necessary	Never Used	Comments
211								X							Unknown - City does not get directly involved in industrial pretreatment.
212								X							Not applicable.
213															Not applicable.
215								X							
216								X				X	X		Bypass overflow used for system maintenance and system failures (infrequent) - item for future growth.
217					X		X				X				
218															Industrial Pretreatment: The utility has a separate pretreatment section. Each pretreatment customer is responsible for their own system. Pretreatment flows are regularly monitored for compliance.
219								X					X		
221															
222															
224															
225															
228															
229															
230															
231															
232															
233															We are not involved. Pretreatment is the responsibilty of the industry.

Table F-3. Reasons for Redundancy, Redundancy Ratings, and General Comments Industrial Pretreatment (continued).

ID	Reasons for Redundancy (Federal, State/Provincial, or Regional Regulations)										Redundancy Ratings				Comments
Plant ID No.	General Pretreatment Regulations	In-House Pollution Prevention Program	Discharge Permit Regulations	Safety Regulations	In-House Operations & Maintenance Requirements	IEEE Standard 493	Emergency/Disaster Planning Regulations/ Requirements	None	Other	Specify	Extremely Effective	Somewhat Effective	Currently Not Necessary	Never Used	Comments
234									X	10-States Standard					No redundant systems for pretreatment.
235									X	10-States Standard					No redundant systems for industrial pretreatment.
236															
237															Our industrial users with pretreatment can generally halt and contain discharge in the case of equipment failure.
238															
239		X	X	X	X		X				X				None
240								X					X		
241								X							No Redundancy.
243															
244	X		X	X											Codes and regulations do not specifically require redundancy. The design and operation of pretreatment systems vary widely and it is largely left to the industry to determine how to best meet the discharge limits and requirements.
245															
246															Pretreatment systems are not operated by TMUA. Permits are issued to industries that require their compliance to Ordiance. Compliance with Ordiance is confirmed based on inspections, self monitoring reports and periodic sampling of effluent.
247															
Total	1	2	5	2	7	0	5	9	2		5	3	3	1	

APPENDIX G

PRELIMINARY TREATMENT

Reference survey data is provided in a tabular format in this appendix for treatment facility preliminary treatment processes. The following tables are included:

- Table G-1. Infrastructure and Equipment
- Table G-2. Instrumentation, Automation and Control
- Table G-3. Reasons for Redundancy, Redundancy Ratings, and General Comments

Key Code:

Safety Regulations	*Occupational Health and Safety Administration—OSHA; National Fire Protection Association—NFPA; Uniform Fire Code—UFC; National Institute of Occupational Safety & Health—NIOSH*
IEEE Standard 493	Institute of Electrical and Electronic Engineers recommended practice for reliable electrical system design
Extremely Effective	Consistently use redundant systems and have been able to reduce failures with these practices
Somewhat Effective	Occasionally use redundant systems for preventative maintenance not for system failures
Currently Not Necessary	The redundant systems of this treatment process area is not currently used but may be used with future growth requirements
Never Used	The redundant systems have never been used and do not foresee a future use

Table G-1. Infrastructure and Equipment Preliminary Treatment.

ID	Infrastructure													Equipment											
	Bypass Units							Backup/Secondary or Multipurpose Units																	
Plant ID No.	Bypass Channel for Screening Units	Grit Removal Bypass Line/ channel	Septage Receiving Station Bypass	Bypass Lines for Flow Equalization Basin/Tanks	None	Other	Specify	Screening and Grit Removal for Septage Receiving	Additional Storage for Septage Receiving	Flow Equalization Basin/ Tanks	None	Other	Specify	Grinders or Comminutors	Bar Screens	Grit Chambers or Hoppers	Aerators in Grit Removal Basin/Tank	Aerators in Septage Receiving Basin/Tank	Septage Receiving Station	Flow Equalization Chambers	Flow Equalization Pumps	Odor Control System	None	Other	Specify
4					X						X				X										
7					X						X				X										
9	X	X													X					X		X			
13					X			X							X										
19				X											X					X	X				
21		X									X														
35	X	X																					X		
44		X							X					X		X									
49				X				X		X				X	X	X	X					X			
60					X					X				X	X										
203					X						X			X	X	X			X						
205		X												X	X	X	X		X			X			
208	X	X												X	X	X	X								
209	X	X									X				X	X									
210	X										X												X		
211		X										X	Take one train out of service in low flows											X	Take one train out of service in low flows
212					X						X				X	X									
213					X							X	Redundant grit tanks		X	X									
215	X	X						X						X	X	X	X								

Table G-1. Infrastructure and Equipment Preliminary Treatment (continued).

ID	Infrastructure													Equipment											
	Bypass Units							Backup/Secondary or Multipurpose Units																	
Plant ID No.	Bypass Channel for Screening Units	Grit Removal Bypass Line/ channel	Septage Receiving Station Bypass	Bypass Lines for Flow Equalization Basin/Tanks	None	Other	Specify	Screening and Grit Removal for Septage Receiving	Additional Storage for Septage Receiving	Flow Equalization Basin/ Tanks	None	Other	Specify	Grinders or Comminutors	Bar Screens	Grit Chambers or Hoppers	Aerators in Grit Removal Basin/Tank	Aerators in Septage Receiving Basin/Tank	Septage Receiving Station	Flow Equalization Chambers	Flow Equalization Pumps	Odor Control System	None	Other	Specify
216					X				X														X		
217					X			X	X						X	X			X						
218	X	X								X					X	X			X	X	X	X			
219																							X		
221											X			X	X	X									
222	X	X									X				X	X									
224					X			X							X	X									
225	X	X						X							X	X	X								
228		X									X				X										
229					X			X							X	X									
230	X										X				X	X						X			
231					X						X				X	X						X			
232					X						X				X	X	X							X	Lift pumps
233	X								X														X		
234					X						X				X	X									
235					X						X				X	X									
236			X								X				X	X						X			
237	X	X													X		X								
238	X	X													X	X									
239	X														X				X			X			

Table G-1. Infrastructure and Equipment Preliminary Treatment (continued).

ID	Infrastructure													Equipment											
	Bypass Units							Backup/Secondary or Multipurpose Units																	
Plant ID No.	Bypass Channel for Screening Units	Grit Removal Bypass Line/ channel	Septage Receiving Station Bypass	Bypass Lines for Flow Equalization Basin/Tanks	None	Other	Specify	Screening and Grit Removal for Septage Receiving	Additional Storage for Septage Receiving	Flow Equalization Basin/ Tanks	None	Other	Specify	Grinders or Comminutors	Bar Screens	Grit Chambers or Hoppers	Aerators in Grit Removal Basin/Tank	Aerators in Septage Receiving Basin/Tank	Septage Receiving Station	Flow Equalization Chambers	Flow Equalization Pumps	Odor Control System	None	Other	Specify
240	X							X							X	X									
241					X						X			X	X										
243	X	X	X	X						X					X	X	X					X			
244					X				X														X		
245		X													X	X									
246					X							X	Redundant Units		X	X	X								
247					X							X	Multiple Units	X	X	X						X			
Total	**16**	**17**	**2**	**3**	**19**	**0**		**8**	**5**	**4**	**17**	**4**		**10**	**37**	**28**	**9**	**0**	**5**	**3**	**2**	**10**	**6**	**2**	

Table G-2. Instrumentation, Automation and Control Preliminary Treatment.

ID	Instrumentation (Backup/Secondary or Multipurpose Units)									Automation and Control (Backup/Secondary or Multipurpose Units)								
Plant ID No.	Level Sensor in Bar Screen Channels	Level Sensor in Septage Receiving Basin/Tank	Level Sensor in Flow Equalization Basin/Tank	Air Flowmeter in Grit Chambers	pH Analyzer in Septage Receiving Station	Flowmeters for Flow Equalization Basin/Tank	None	Other	Specify	Dual Power Source	Dual PLCs	Portable or In-Place Generators	Multiple Control Strategies	DCS/SCADA System Including Telemetry	Dual Field-Mounted Control Boxes	None	Other	Specify
4					X									X				
7							X									X		
9	X									X		X	X	X				
13							X			X	X		X	X				
19	X					X				X		X	X	X				
21																		
35							X			X								
44							X			X				X				
49	X	X		X	X							X		X				
60										X								UPS
203	X									X		X	X	X				
205	X			X						X		X	X	X				
208	X	X												X				
209							X					X						
210							X					X						
211							X							X			X	Dual network loop
212							X			X								
213							X			X		X		X				
215	X			X								X	X	X				
216							X									X		
217							X			X	X							
218	X		X											X				
219								X	explosive gas meter	X	X	X			X			

Table G-2. Instrumentation, Automation and Control Preliminary Treatment (continued).

ID	Instrumentation (Backup/Secondary or Multipurpose Units)									Automation and Control (Backup/Secondary or Multipurpose Units)								
Plant ID No.	Level Sensor in Bar Screen Channels	Level Sensor in Septage Receiving Basin/Tank	Level Sensor in Flow Equalization Basin/Tank	Air Flowmeter in Grit Chambers	pH Analyzer in Septage Receiving Station	Flowmeters for Flow Equalization Basin/Tank	None	Other	Specify	Dual Power Source	Dual PLCs	Portable or In-Place Generators	Multiple Control Strategies	DCS/SCADA System Including Telemetry	Dual Field-Mounted Control Boxes	None	Other	Specify
									portable pH, D.O. meter, temp and timer control to back up bar screen									
221							X			X		X						
222							X			X	X	X	X	X				
224							X			X			X					
225							X			X			X	X				
228	X									X			X					
229							X			X		X	X					
230							X			X		X	X	X				
231							X			X		X	X	X				
232							X									X		
233							X			X							X	The treatment plant has a dual power source.
234							X			X							X	10-States Standard
235							X										X	10-States Standard
236	X									X			X	X				
237							X			X								
238								X	Uninterruptible power supplies (AC with battery backups)	X				X			X	Uninterruptible power supplies (AC with battery backups)
239		X			X					X	X	X	X	X	X			
240							X			X			X	X				
241							X									X		

Table G-2. Instrumentation, Automation and Control Preliminary Treatment (continued).

ID	Instrumentation (Backup/Secondary or Multipurpose Units)									Automation and Control (Backup/Secondary or Multipurpose Units)								
Plant ID No.	Level Sensor in Bar Screen Channels	Level Sensor in Septage Receiving Basin/Tank	Level Sensor in Flow Equalization Basin/Tank	Air Flowmeter in Grit Chambers	pH Analyzer in Septage Receiving Station	Flowmeters for Flow Equalization Basin/Tank	None	Other	Specify	Dual Power Source	Dual PLCs	Portable or In-Place Generators	Multiple Control Strategies	DCS/SCADA System Including Telemetry	Dual Field-Mounted Control Boxes	None	Other	Specify
243	X		X	X						X	X	X		X				
244	X									X		X		X	X			
245										X			X	X				
246	X									X			X	X				
247	X									X		X		X	X			
Total	14	3	2	4	3	1	25	2		32	6	19	18	26	4	4	5	

Table G-3. Reasons for Redundancy, Redundancy Ratings, and General Comments Preliminary Treatment (grinding, screening, grit removal, septage, flow equalization, etc.).

ID	Reasons for Redundancy (Federal, State/Provincial, or Regional Regulations, etc.)								Redundancy Ratings				Preliminary Treatment Comments
Plant ID No.	Septage Treatment and Disposal Regulations	Safety Regulations	In-House Operations and Maintenance Requirements	IEEE Standard 493	Emergency/Disaster Planning Regulations/ Requirements	None	Other	Specify	Extremely Effective	Somewhat Effective	Currently Not Necessary	Never Used	Comments
4			X						X				pH cross checks with multiple meters.
7			X		X								
9			X		X				X				
13			X						X				
19					X					X			
21													
35			X								X		
44			X							X			
49									X				
60			X						X				
203		X	X				X	NFPA 820	X				
205			X						X				
208	X		X		X					X			When necessary to take one unit out of service we're capable of operating with lesser units for a short duration of time.
209	X									X			
210						X							No formal redundacy practices in place.
211			X							X			
212			X		X				X				
213			X						X				
215			X				X	Water Resources Act	X				

Water Environment Research Foundation

Table G-3. Reasons for Redundancy, Redundancy Ratings, and General Comments Preliminary Treatment (grinding, screening, grit removal, septage, flow equalization, etc.) (continued).

ID	Reasons for Redundancy (Federal, State/Provincial, or Regional Regulations, etc.)								Redundancy Ratings				Preliminary Treatment Comments
Plant ID No.	Septage Treatment and Disposal Regulations	Safety Regulations	In-House Operations and Maintenance Requirements	IEEE Standard 493	Emergency/Disaster Planning Regulations/ Requirements	None	Other	Specify	Extremely Effective	Somewhat Effective	Currently Not Necessary	Never Used	Comments
216		X					X	RCRA	X				
217			X		X					X			
218					X						X		
219		X	X	X	X				X				
221			X		X				X				
222		X	X		X					X			
224			X		X					X			
225			X		X					X			
228			X							X			
229		X	X		X					X			
230		X	X		X					X			
231			X						X				
232			X				X	State Regulation					
233			X							X			
234							X	10-States Standard	X				
235							X	10-States Standard	X				
236			X						X				
237			X				X	Federal, state and 10-state stds. for plant power redundancy		X			
238			X						X				
239	X	X	X		X				X				NONE!!!

Table G-3. Reasons for Redundancy, Redundancy Ratings, and General Comments Preliminary Treatment (grinding, screening, grit removal, septage, flow equalization, etc.) (continued).

ID	Reasons for Redundancy (Federal, State/Provincial, or Regional Regulations, etc.)								Redundancy Ratings				Preliminary Treatment Comments
Plant ID No.	Septage Treatment and Disposal Regulations	Safety Regulations	In-House Operations and Maintenance Requirements	IEEE Standard 493	Emergency/Disaster Planning Regulations/ Requirements	None	Other	Specify	Extremely Effective	Somewhat Effective	Currently Not Necessary	Never Used	Comments
240		X	X	X	X				X				
241						X				X			
243		X	X	X	X					X			
244			X						X				
245			X		X					X			
246			X						X				
247			X				X	State Standards	X				
Total	**3**	**9**	**36**	**3**	**18**	**2**	**8**		**23**	**17**	**2**	**0**	

APPENDIX H

PRIMARY TREATMENT

Reference survey data is provided in a tabular format in this appendix for treatment facility primary treatment processes. The following tables are included:

- Table H-1. Infrastructure and Equipment
- Table H-2. Instrumentation, Automation and Control
- Table H-3. Reasons for Redundancy, Redundancy Ratings, and General Comments

Key Code:

Safety Regulations	*Occupational Health and Safety Administration—OSHA; National Fire Protection Association—NFPA; Uniform Fire Code—UFC; National Institute of Occupational Safety & Health—NIOSH*
IEEE Standard 493	Institute of Electrical and Electronic Engineers recommended practice for reliable electrical system design
Extremely Effective	Consistently use redundant systems and have been able to reduce failures with these practices
Somewhat Effective	Occasionally use redundant systems for preventative maintenance not for system failures
Currently Not Necessary	The redundant systems of this treatment process area is not currently used but may be used with future growth requirements
Never Used	The redundant systems have never been used and do not foresee a future use

Table H-1. Infrastructure and Equipment Primary Treatment (skimming, sedimentation, etc.).

ID	Infrastructure								Equipment					
	Bypass Units				Backup/Secondary or Multipurpose Units									
Plant ID No.	Bypass for Sedimentation/ Settling Tank	None	Other	Specify	Sedimentation/ Settling Tank	None	Other	Specify	Skimmers	Scum Pumps	Sludge Pumps	None	Other	Specify
4		X				X						X		
7		X				X					X			
9	X				X					X	X			
19		X			X				X	X	X			
21										X	X			
35	X										X			
44	X				X					X	X			
49		X			X				X	X	X			
60														
203		X								X	X			
205					X				X	X	X			
208		X				X								
209	X				X					X	X			
210		X				X					X			
211	X						X	Take one train out of service in low flows. (winter/summer)					X	Take one train out of service in low flows. (winter/summer)
212	X				X					X	X			
213		X			X							X		
215	X				X				X	X	X			
216		X				X						X		
217		X				X					X			
218		X				X								
219		X			X				X	X	X			
221					X					X	X			
222	X				X						X			

Table H-1. Infrastructure and Equipment Primary Treatment (skimming, sedimentation, etc.) (continued).

ID	Infrastructure								Equipment					
	Bypass Units				Backup/Secondary or Multipurpose Units									
Plant ID No.	Bypass for Sedimentation/ Settling Tank	None	Other	Specify	Sedimentation/ Settling Tank	None	Other	Specify	Skimmers	Scum Pumps	Sludge Pumps	None	Other	Specify
224		X			X					X	X			
225	X				X					X	X			
228		X				X					X			
229														
230		X			X					X	X			
231		X			X					X	X		X	chemical pumps - polymer system
232		X			X				X	X	X			
233		X				X					X			
234		X				X				X	X			
235		X				X				X	X			
236		X			X					X	X			
237		X									X			
238					X					X	X			
239				Bypass holding tanks only		X			X	X	X			
240		X				X								
241		X				X				X	X			
243	X				X					X	X			
244	X				X							X		
245		X				X				X	X			
246		X					X	Multiple Units		X	X			
247		X			X					X	X			
Total	**11**	**27**	**0**		**22**	**15**	**2**		**7**	**27**	**35**	**4**	**2**	

Table H-2. Instrumentation, Automation and Control Primary Treatment (skimming, sedimentation, etc.).

ID	Instrumentation							Automation and Control								
Plant ID No.	Water Level Sensors	Loss of Flow Sensors	Interface/Sludge Blanket Level	Suspended Solids/ Sludge Density Meters	None	Other	Specify	Dual Power Source	Dual PLCs	Portable or In-Place Generators	Multiple Control Strategies	DCS/SCADA System Including Telemetry	Dual Field-Mounted Control Boxes	None	Other	Specify
4					X							X				
7					X									X		
9								X		X	X	X				
19					X			X		X		X				
21					X					X						
35	X											X				
44				X				X				X				
49	X		X							X		X				
60																
203					X			X			X	X			X	Polymer addition system
205	X		X					X		X	X	X				
208					X							X				
209	X	X								X						
210					X					X						
211					X							X			X	Dual network loop.
212					X									X		
213					X			X		X		X				
215	X			X						X	X	X				
216					X									X		
217					X			X								
218					X									X		
219					X			X		X	X		X			
221								X								
222					X			X	X	X	X	X				

Table H-2. Instrumentation, Automation and Control Primary Treatment (skimming, sedimentation, etc.) (continued).

ID	Instrumentation							Automation and Control								
Plant ID No.	Water Level Sensors	Loss of Flow Sensors	Interface/Sludge Blanket Level	Suspended Solids/ Sludge Density Meters	None	Other	Specify	Dual Power Source	Dual PLCs	Portable or In-Place Generators	Multiple Control Strategies	DCS/SCADA System Including Telemetry	Dual Field-Mounted Control Boxes	None	Other	Specify
224					X			X			X					
225					X			X			X	X				
228					X			X			X					
229																
230			X					X		X	X	X				
231					X			X		X	X	X				
232					X									X		
233					X			X								
234					X			X							X	10-States Standard
235					X										X	10-States Standard
236	X							X		X	X	X				
237					X			X								
238								X				X			X	Uninterruptible power supplies (AC with battery backups)
239	X	X	X					X	X	X	X	X	X			
240					X									X		
241					X									X		
243	X	X		X				X	X	X		X				
244	X							X		X		X	X			
245								X			X	X				
246					X			X			X	X				
247	X		X					X		X		X	X			
Total	**10**	**3**	**5**	**3**	**27**	**0**		**26**	**3**	**18**	**15**	**24**	**4**	**7**	**5**	

Table H-3. Reasons for Redundancy, Redundancy Ratings, and General Comments Primary Treatment (skimming, sedimentation, etc.).

ID	Reasons for Redundancy (Federal, State/Provincial, or Regional Regulations, etc.)								Redundancy Ratings				Comments
Plant ID No.	Discharge Permit Regulations	Safety Regulations	In-House Operations and Maintenance Requirements	IEEE Standard 493	Emergency/Disaster Planning Regulations/Requirements	None	Other	Specify	Extremely Effective	Somewhat Effective	Currently Not Necessary	Never Used	Comments
4			X										
7			X		X								
9			X		X				X				
19	X				X					X			We have 8 primary clarifiers of identical dimensions and capacities, supported by multiple scum and sludge pumps and interconnected piping.
21			X							X			
35	X									X			
44			X							X			
49	X		X		X				X				
60													
203		X	X				X	NFPA 820	X				
205	X		X						X				
208	X		X		X					X			7 clarifiers - the plant can operate with 6 if necessary- using 7 online improves effluent quality. When necessary to take one unit out of service we're capable of operating with lesser units for a short duration of time.
209										X			
210						X							No formal redundacy practices in place.
211			X							X			
212	X		X		X				X				
213			X						X				
215			X				X	Water Resources Act	X				
216						X							Not used.

Table H-3. Reasons for Redundancy, Redundancy Ratings, and General Comments Primary Treatment (skimming, sedimentation, etc.) (continued).

ID	Reasons for Redundancy (*Federal, State/Provincial, or Regional Regulations, etc.*)								Redundancy Ratings				Comments
Plant ID No.	Discharge Permit Regulations	Safety Regulations	In-House Operations and Maintenance Requirements	IEEE Standard 493	Emergency/Disaster Planning Regulations/Requirements	None	Other	Specify	Extremely Effective	Somewhat Effective	Currently Not Necessary	Never Used	Comments
217			X		X					X			
218						X						X	
219	X		X	X					X				
221	X		X		X				X				
222		X	X		X					X			
224			X		X					X			
225	X		X		X					X			
228	X		X							X			
229													
230		X	X		X					X			
231			X						X				
232			X							X			
233			X							X			
234							X	10-States Standard	X				
235							X	10-States Standard	X				
236	X				X				X				
237	X		X				X	Federal, state and 10-state stds. for plant power redundancy	X				Rating based on dual power source redundancy for operation during power outage on one source
238			X						X				
239	X	X	X		X				X				NONE !!!!!!!
240						X						X	
241						X				X			

Table H-3. Reasons for Redundancy, Redundancy Ratings, and General Comments Primary Treatment (skimming, sedimentation, etc.) (continued).

ID	Reasons for Redundancy (*Federal, State/Provincial, or Regional Regulations, etc.*)								Redundancy Ratings				Comments
Plant ID No.	Discharge Permit Regulations	Safety Regulations	In-House Operations and Maintenance Requirements	IEEE Standard 493	Emergency/Disaster Planning Regulations/Requirements	None	Other	Specify	Extremely Effective	Somewhat Effective	Currently Not Necessary	Never Used	Comments
243	X	X	X	X	X					X			
244	X		X						X				
245			X							X			
246			X						X				
247			X				X	State Standards	X				
Total	**15**	**5**	**32**	**2**	**15**	**5**	**6**		**19**	**18**	**0**	**2**	

SECONDARY TREATMENT

Reference survey data is provided in a tabular format in this appendix for treatment facility secondary treatment processes. The following tables are included:

- Table I-1. Infrastructure and Equipment
- Table I-2. Instrumentation, Automation and Control
- Table I-3. Reasons for Redundancy, Redundancy Ratings, and General Comments

Key Code:

Safety Regulations	*Occupational Health and Safety Administration—OSHA; National Fire Protection Association—NFPA; Uniform Fire Code—UFC; National Institute of Occupational Safety & Health—NIOSH*
IEEE Standard 493	Institute of Electrical and Electronic Engineers recommended practice for reliable electrical system design
Extremely Effective	Consistently use redundant systems and have been able to reduce failures with these practices
Somewhat Effective	Occasionally use redundant systems for preventative maintenance not for system failures
Currently Not Necessary	The redundant systems of this treatment process area is not currently used but may be used with future growth requirements
Never Used	The redundant systems have never been used and do not foresee a future use

Table I-1. Infrastructure and Equipment Secondary Treatment (attached growth, suspended growth, etc.).

ID	Infrastructure												Equipment									
	Bypass Units					Backup/Secondary or Multipurpose Units																
Plant ID No.	Sludge Bypass Channels or Piping	Oxygen Bypass Line	None	Other	Specify	Secondary Process Retention Basin/Tanks	Suspended Growth Basin Tanks	Attached Growth Basin/ Tanks	Oxygen Storage Tanks	None	Other	Specify	RAS/WAS Pumping	Sludge Pumping	Mixers	Aerators	Blowers	Oxygen Compressors	Chlorinators	None	Other	Specify
4			X			X								X								
7			X							X				X								
9	X									X			X	X	X							
19			X				X						X	X	X		X					
21			X			X							X	X	X	X	X					
35	X																					
44			X										X	X		X	X		X			
49			X			X	X	X					X	X		X	X					
60			X				X						X	X			X					
203			X				X						X				X	X				
205	X	X				X	X		X				X		X							
208	X			X	Bypass return activated sludge line					X			X	X			X		X			
209							X							X			X					
210	X									X			X				X					
211				X	Take one train out of service in low flows. (winter/summer)				X		X	Take one train out of service in low flows. (winter/summer)									X	Take one train out of service in low flows. (winter/ summer)
212				X	Influent bypass	X							X		X							
213			X				X	X					X				X					
215				X	One effluent bypass					X			X				X		X			
216			X							X			X			X	X				X	process effluent pumps

Table I-1. Infrastructure and Equipment Secondary Treatment (attached growth, suspended growth, etc.) (continued).

ID	Infrastructure												Equipment									
	Bypass Units					Backup/Secondary or Multipurpose Units																
Plant ID No.	Sludge Bypass Channels or Piping	Oxygen Bypass Line	None	Other	Specify	Secondary Process Retention Basin/Tanks	Suspended Growth Basin Tanks	Attached Growth Basin/Tanks	Oxygen Storage Tanks	None	Other	Specify	RAS/WAS Pumping	Sludge Pumping	Mixers	Aerators	Blowers	Oxygen Compressors	Chlorinators	None	Other	Specify
217			X							X			X	X								
218	X						X						X	X			X		X			
219			X				X										X					
221							X						X				X		X			
222			X				X						X	X			X					
224	X						X							X			X		X			
225	X					X	X						X	X			X					
228			X							X			X	X			X					
229	X									X			X	X			X		X			
230	X						X						X	X			X					
231																						
232				X	2 Oxygen plants interconnected	X	X		X				X	X	X	X	X	X	X			
233			X							X			X				X					
234			X				X				X	Clarifiers	X				X					
235			X			X					X	Clarifiers	X				X					
236	X						X						X				X					
237					Wet weather high flow bypass														X			multi-purpose influent feed pumps
238		X				X	X		X				X	X			X	X	X			
239			X			X							X	X	X	X	X					
240		X							X				X	X		X					X	Scum system

Table I-1. Infrastructure and Equipment Secondary Treatment (attached growth, suspended growth, etc.) (continued).

ID	Infrastructure												Equipment									
	Bypass Units					Backup/Secondary or Multipurpose Units																
Plant ID No.	Sludge Bypass Channels or Piping	Oxygen Bypass Line	None	Other	Specify	Secondary Process Retention Basin/Tanks	Suspended Growth Basin Tanks	Attached Growth Basin/ Tanks	Oxygen Storage Tanks	None	Other	Specify	RAS/WAS Pumping	Sludge Pumping	Mixers	Aerators	Blowers	Oxygen Compressors	Chlorinators	None	Other	Specify
241							X		X				X	X	X		X	X	X			
243			X				X						X	X	X		X					
244							X		X				X	X	X	X		X	X			
245	X												X	X			X					
246			X								X	Multiple Units	X	X		X			X			
247						X							X	X	X		X		X			
Total	**12**	**3**	**20**	**5**		**11**	**21**	**2**	**7**	**10**	**4**		**36**	**28**	**11**	**9**	**3 2**	**5**	**14**	**0**	**3**	

Water Environment Research Foundation

Table I-2. Instrumentation, Automation and Control Secondary Treatment (attached growth, suspended growth, etc.).

ID	Instrumentation										Automation and Control								
Plant ID No.	Dissolved Oxygen Analyzers	Flowmeters (RAS/ WAS)	Air Flowmeter for Blowers	Interface/Sludge Blanket Level	Water Level Sensors	Suspended Solids Analyzers	Oxygen Purity Analyzers	None	Other	Specify	Dual Power Source	Dual PLCs	Portable or In-Place Generators	Multiple Control Strategies	DCS/SCADA System Including Telemetry	Dual Field-Mounted Control Boxes	None	Other	Specify
4								X							X				
7								X									X		
9	X										X		X	X	X				
19	X								X	Biologic Activity Meters for Nitrification and Denitrification	X		X	X	X				
21								X							X				
35	X		X								X				X				
44	X	X									X				X				
49	X	X	X										X		X				
60																		X	UPS
203	X				X	X					X			X	X				
205	X	X	X		X		X			Airflow meters are for the cryogenic plant	X		X	X	X				
208	X												X		X				
209								X					X						
210								X					X		X				
211									X	May have spare units available on occasion.		X		X	X				
212								X						X					
213	X	X									X		X		X				
215	X	X		X		X								X					
216								X									X		
217	X																		
218	X	X	X			X									X				
219	X										X	X		X		X			
221								X			X								
222	X			X		X					X	X	X	X	X				

Table I-2. Instrumentation, Automation and Control Secondary Treatment (attached growth, suspended growth, etc.) (continued).

ID	Instrumentation										Automation and Control								
Plant ID No.	Dissolved Oxygen Analyzers	Flowmeters (RAS/ WAS)	Air Flowmeter for Blowers	Interface/Sludge Blanket Level	Water Level Sensors	Suspended Solids Analyzers	Oxygen Purity Analyzers	None	Other	Specify	Dual Power Source	Dual PLCs	Portable or In-Place Generators	Multiple Control Strategies	DCS/SCADA System Including Telemetry	Dual Field-Mounted Control Boxes	None	Other	Specify
224	X										X			X					
225								X			X			X	X				
228	X		X			X					X			X					
229								X			X		X	X					
230				X							X		X		X				
231																			
232	X						X				X					X			
233								X			X			X					
234								X			X								
235								X			X								
236		X									X			X	X				
237					X						X								
238											X				X			X	Uninterruptible power supplies (AC with battery backups)
239	X	X	X	X							X	X		X	X	X			
240								X			X			X	X				
241								X									X		
243	X	X	X		X						X	X	X		X				
244	X	X	X	X		X	X				X			X	X				
245				X							X			X	X				
246	X										X		X	X	X				
247	X		X	X							X			X	X	X			
Total	**22**	**10**	**9**	**7**	**4**	**6**	**3**	**15**	**2**		**29**	**5**	**13**	**21**	**26**	**4**	**3**	**2**	

Table I-3. Reasons for Redundancy, Redundancy Ratings, and General Comments Secondary Treatment (attached growth, suspended growth, etc.) .

ID	Reasons for Redundancy *(Federal, State/Provincial, or Regional Regulations, etc.)*								Redundancy Ratings				
Plant ID No.	Discharge Permit Regulations	Safety Regulations	In-House Operations & Maintenance Requirements	IEEE Standard 493	Emergency/Disaster Planning Regulations/ Requirements	None	Other	Specify	Extremely Effective	Somewhat Effective	Currently Not Necessary	Never Used	Comments
4						X							
7			X		X				X				
9			X		X				X				
19	X				X					X			
21			X							X			
35			X						X				
44			X							X			
49	X		X		X				X				
60			X						X				
203	X	X	X				X	NFPA 820	X				
205	X		X						X				
208	X	X	X		X				X				
209			X							X			
210						X							No formal redundancy practices in place.
211			X							X			
212			X							X			
213			X						X				
215			X				X	Water Resources Act	X				
216						X				X			
217			X						X				
218					X						X		
219			X				X	10-State Standards	X				
221	X		X		X				X				
222		X	X		X					X			
224			X		X					X			

Table I-3. Reasons for Redundancy, Redundancy Ratings, and General Comments Secondary Treatment (attached growth, suspended growth, etc.) (continued).

ID	Reasons for Redundancy *(Federal, State/Provincial, or Regional Regulations, etc.)*								Redundancy Ratings				
Plant ID No.	Discharge Permit Regulations	Safety Regulations	In-House Operations & Maintenance Requirements	IEEE Standard 493	Emergency/Disaster Planning Regulations/ Requirements	None	Other	Specify	Extremely Effective	Somewhat Effective	Currently Not Necessary	Never Used	Comments
225	X		X		X					X			
228	X		X						X				
229		X	X		X					X			
230		X	X		X					X			
231													
232			X							X			
233			X							X			
234							X	Good Design Practices	X				
235							X	Good Design Practices	X				
236	X		X							X			
237	X							Federal, state and 10-state stds. for plant power redundancy	X				
238			X						X				Redundant storage tanks for LOX.
239	X	X	X		X					X			NONE!!!!!!!!!!!!
240	X		X		X				X				
241			X		X				X				
243	X	X	X	X	X					X			
244	X								X				
245			X							X			
246			X						X				
247			X				X	State Standards	X				
Total	**14**	**7**	**35**	**1**	**16**	**3**	**6**		**23**	**18**	**1**	**0**	

TERTIARY/ADVANCED TREATMENT

Reference survey data is provided in a tabular format in this appendix for treatment facility tertiary/advanced treatment processes. The following tables are included:

- Table J-1. Infrastructure and Equipment
- Table J-2. Instrumentation, Automation and Control
- Table J-3. Reasons for Redundancy, Redundancy Ratings, and General Comments

Key Code:

Safety Regulations	*Occupational Health and Safety Administration—OSHA; National Fire Protection Association—NFPA; Uniform Fire Code—UFC; National Institute of Occupational Safety & Health—NIOSH*
IEEE Standard 493	Institute of Electrical and Electronic Engineers recommended practice for reliable electrical system design
Extremely Effective	Consistently use redundant systems and have been able to reduce failures with these practices
Somewhat Effective	Occasionally use redundant systems for preventative maintenance not for system failures
Currently Not Necessary	The redundant systems of this treatment process area is not currently used but may be used with future growth requirements
Never Used	The redundant systems have never been used and do not foresee a future use

Table J-1. Infrastructure and Equipment Tertiary/Advanced Treatment (filtration, nutrient removal, etc.).

ID	Infrastructure											Equipment (*Backup/Secondary or Multipurpose Units*)									
	Bypass Units					Backup/Secondary or Multipurpose Units															
Plant ID No.	Bypass Channel for Secondaries	Bypass Filtration	None	Other	Specify	Nutrient Removal Basin/Tank	Filters	Backwash Tank	None	Other	Specify	Sludge Pumps	Filter Pumps	Backwash Pumps	Filter Media	Aerators	Baffles	Odor Control System	None	Other	Specify
4			X						X										X		
7			X						X										X		
9	X					X						X									
19		X				X	X	X				X	X	X							
21																					
35																					
44	X	X							X			X	X	X							
49			X						X			X			X	X		X			
60		X					X												X		
203																					
205																					
208		X					X	X		X	Operating under design capacity allows us to bypass tanks on cells	X	X	X	X						
209	X					X						X									
210	X	X							X					X							
211			X						X										X		
212		X							X			X		X							
213																					
215			X						X										X		
216	X									X	Wetlands cell								X		
217			X						X										X		
218		X							X										X		
219	X	X				X	X							X							
221			X						X										X		

Table J-1. Infrastructure and Equipment Tertiary/Advanced Treatment (filtration, nutrient removal, etc.) (continued).

ID	Infrastructure											Equipment *(Backup/Secondary or Multipurpose Units)*									
	Bypass Units					Backup/Secondary or Multipurpose Units															
Plant ID No.	Bypass Channel for Secondaries	Bypass Filtration	None	Other	Specify	Nutrient Removal Basin/Tank	Filters	Backwash Tank	None	Other	Specify	Sludge Pumps	Filter Pumps	Backwash Pumps	Filter Media	Aerators	Baffles	Odor Control System	None	Other	Specify
222																					
224		X					X												X		
225		X					X						X	X	X						
228																					
229		X							X			X	X	X							
230																					
231																					
232				X	Not Applicable					X	Not Applicable										
233			X						X										X		
234			X						X										X		
235																					
236																					
237	X												X								
238																					
239		X										X	X	X	X						
240		X					X							X	X					X	waste pumps
241																					
243							X						X	X	X						
244	X						X	X				X	X	X				X			
245																					
246																					
247										X	multiple units	X	X	X							
Total	**8**	**13**	**9**	**1**		**4**	**9**	**3**	**14**	**4**		**11**	**10**	**13**	**6**	**1**	**0**	**2**	**12**	**1**	

Table J-2. Instrumentation, Automation and Control Tertiary/Advanced Treatment (filtration, nutrient removal, etc.).

ID	Instrumentation														Automation and Control								
Plant ID No.	Level Sensors	Ammonia Analyzer	Nitrate/Nitrite Analyzer	Phosphorus Analyzer	pH Analyzer	Dissolved Oxygen Analyzer	BOD Analyzer	COD Analyzer	Respirometer	Flowmeters	ORP Meter	None	Other	Specify	Dual Power Source	Dual PLCs	Portable or In-Place Generators	Multiple Control Strategies	DCS/SCADA System Including Telemetry	Dual Field-Mounted Control Boxes	None	Other	Specify
4												X							X				
7	X																		X				
9						X									X		X	X	X				
19	X	X	X			X								Turbidity meter	X		X	X	X				
21																							
35																							
44	X				X										X				X				
49					X	X											X		X				
60												X										X	UPS
203																							
205																							
208						X											X		X				
209												X					X						
210												X					X		X				
211												X									X		
212												X						X					
213																							
215																							
216												X									X		
217												X									X		
218												X							X				
219					X	X							X	alkalinity testing	X		X	X		X			
221						X									X								
222																							
224												X			X			X					

Table J-2. Instrumentation, Automation and Control Tertiary/Advanced Treatment (filtration, nutrient removal, etc.) (continued).

ID	Instrumentation														Automation and Control								
Plant ID No.	Level Sensors	Ammonia Analyzer	Nitrate/Nitrite Analyzer	Phosphorus Analyzer	pH Analyzer	Dissolved Oxygen Analyzer	BOD Analyzer	COD Analyzer	Respirometer	Flowmeters	ORP Meter	None	Other	Specify	Dual Power Source	Dual PLCs	Portable or In-Place Generators	Multiple Control Strategies	DCS/SCADA System Including Telemetry	Dual Field-Mounted Control Boxes	None	Other	Specify
225												X			X			X	X				
228																							
229												X			X		X	X					
230																							
231																							
232																							
233												X			X			X					
234												X			X								
235															X								
236																							
237	X														X								
238																							
239	X			X	X	X				X				Portable analyzers	X	X	X	X	X	X			
240												X			X			X	X				
241																							
243	X				X	X				X					X	X	X		X				
244															X			X	X				
245																							
246																							
247	X									X					X		X	X	X	X			
Plant	**7**	**1**	**1**	**1**	**5**	**8**	**0**	**0**	**0**	**3**	**0**	**15**	**1**		**17**	**2**	**11**	**12**	**15**	**3**	**3**	**1**	

Table J-3. Reasons for Redundancy, Redundancy Ratings, and General Comments Tertiary/Advanced Treatment (filtration, nutrient removal, etc.).

ID	Reasons for Redundancy								Redundancy Ratings				Comments
Plant ID No.	Discharge Permit Regulations	Safety Regulations	In-House Operations and Maintenance Requirements	IEEE Standard 493	Emergency/Disaster Planning Regulations/ Requirements	None	Other	Specify	Extremely Effective	Somewhat Effective	Currently Not Necessary	Never Used	Comments
4						X							
7			X		X				X				
9			X		X								
19	X				X					X			
21													
35			X						X				
44			X							X			
49						X			X				
60			X						X				
203													
205													
208	X	X	X		X				X				
209			X							X			
210						X							No formal redundancy practices in place.
211						X							Not applicable. We have no formal tertiary treatment.
212			X						X				
213													Not applicable.
215													
216						X					X		
217						X							
218	X				X						X		
219	X			X					X				
221	X		X		X				X				
222													

Table J-3. Reasons for Redundancy, Redundancy Ratings, and General Comments Tertiary/Advanced Treatment (filtration, nutrient removal, etc.) (continued).

ID	Reasons for Redundancy								Redundancy Ratings				Comments
Plant ID No.	Discharge Permit Regulations	Safety Regulations	In-House Operations and Maintenance Requirements	IEEE Standard 493	Emergency/Disaster Planning Regulations/ Requirements	None	Other	Specify	Extremely Effective	Somewhat Effective	Currently Not Necessary	Never Used	Comments
224			X		X					X			
225	X		X		X					X			
228													
229		X	X		X					X			
230													
231													
232													
233			X							X			We remove nitrogen and phosphorus biologically in the secondary treatment system.
234							X	Good Design Practices	X				
235							X	Good Design Practices	X				
236													
237	X		X					Federal, state and 10-state stds. for plant power redundancy	X				
238													Do not do Tertiary/Advanced Treament
239	X	X	X		X				X				NONE!!!!!!!!!!!!!!!!!!!!!!!!!!!!!!!
240	X								X				
241													Not Applicable
243	X	X	X	X							X		
244			X							X			
245													
246													Not Applicable
247			X				X	State standards	X				
Plant	**10**	**4**	**18**	**2**	**10**	**6**	**3**		**14**	**8**	**3**	**0**	

APPENDIX K

SOLIDS PROCESSING

Reference survey data is provided in a tabular format in this appendix for treatment facility solids processing. The following tables are included:

- Table K-1. Infrastructure and Equipment
- Table K-2. Instrumentation, Automation and Control
- Table K-3. Reasons for Redundancy, Redundancy Ratings, and General Comments

Table K-1. Infrastructure and Equipment Solids Processing (digestion, thickening, dewatering, composting, etc.).

ID	Infrastructure																	Equipment								
	Bypass Units					Backup/Secondary or Multipurpose Units																				
Plant ID No.	Bypass Lines for Solid Processing	Digester Gas Bypass Lines	None	Other	Specify	Digesters	Centrifuges	Gravity Belt Presses	Gravity Belt Thickeners	Plate Filter Presses	Dissolved Air Flotation (DAF)	Digester Gas Holing Tanks	Digester Gas Flares	Chemical Feed Tanks	None	Other	Specify	Sludge Pumps	Gas Compressors	Chemical Feed Pumps	Chemical Mixers	Odor Control System	Boilers/Steam Supply	None	Other	Specify
4			X					X																X		
7			X												X									X		
9		X				X					X	X	X					X	X				X			
19	X							X	X					X				X		X	X					
21			X												X			X								
35	X					X												X								
44			X					X	X									X			X					
49	X						X		X					X				X	X	X	X	X	X			
60															X			X								
203		X				X	X				X		X	X				X		X	X		X			
205	X	X				X	X				X		X	X				X	X	X	X	X	X			
208	X					X		X										X								
209			X			X								X				X	X	X			X			
210		X		X	Flare Stack								X											X		
211			X			X	X							X				X		X			X			
212	X							X										X		X						
213	X					X	X				X		X					X	X				X			
215	X	X				X	X				X							X	X	X	X		X			
216			X												X									X		
217			X					X										X		X	X					
218		X						X				X	X	X				X	X	X						
219	X						X		X							X	thermal conditioning, incineration, gravity thickeners	X		X		X	X			

Water Environment Research Foundation

Table K-1. Infrastructure and Equipment Solids Processing (digestion, thickening, dewatering, composting, etc.) Continued).

ID	Infrastructure																	Equipment								
	Bypass Units					Backup/Secondary or Multipurpose Units																				
Plant ID No.	Bypass Lines for Solid Processing	Digester Gas Bypass Lines	None	Other	Specify	Digesters	Centrifuges	Gravity Belt Presses	Gravity Belt Thickeners	Plate Filter Presses	Dissolved Air Flotation (DAF)	Digester Gas Holing Tanks	Digester Gas Flares	Chemical Feed Tanks	None	Other	Specify	Sludge Pumps	Gas Compressors	Chemical Feed Pumps	Chemical Mixers	Odor Control System	Boilers/Steam Supply	None	Other	Specify
221											X							X								
222	X	X				X	X						X	X				X	X	X		X	X			
224	X	X				X			X				X					X					X			
225	X	X				X	X		X			X	X	X				X	X	X	X		X			
228		X							X									X		X						
229	X														X			X								
230																										
231										X				X				X		X	X	X	X			
232	X						X	X						X				X	X	X	X					
233	X	X													X			X	X	X			X			
234			X			X	X			X	X		X					X	X	X						
235			X						X									X		X	X					
236		X				X		X	X			X						X	X	X	X		X			
237				X	storage													X								
238		X				X			X		X	X	X	X				X	X	X		X	X			
239			X							X	X			X				X		X	X	X				
240	X	X				X	X				X			X				X		X	X				X	Heat exchangers
241			X								X							X					X			
243							X				X			X				X	X	X	X	X	X			
244						X	X	X				X	X					X	X	X	X		X			
245	X	X				X	X						X			X	lagoons	X	X	X			X			
246				X	Digested Biosolids storage lagoon	X		X					X	X				X	X	X	X	X	X			
247			X			X	X				X		X					X	X	X	X	X	X			
Total	**17**	**15**	**13**	**3**		**20**	**16**	**11**	**10**	**3**	**13**	**6**	**15**	**16**	**6**	**2**		**40**	**19**	**28**	**18**	**10**	**22**	**4**	**1**	

Table K-2. Instrumentation, Automation and Control Solids Processing (digestion, thickening, dewatering, composting, etc.).

ID	Instrumentation												Automation and Control								
Plant ID No.	Suspended Solids Analyzers	Sludge Flowmeters	Chemical Feed Flowmeters	Gas Flowmeters	Temperature Sensors	Gas Detectors	Fire Sensors	Level Sensors	Solids Volume, Weight and Concentration Meters	None	Other	Specify	Dual Power Source	Dual PLCs	Portable or In-Place Generators	Multiple Control Strategies	DCS/SCADA System Including Telemetry	Dual Field-Mounted Control Boxes	None	Other	Specify
217										X									X		
218				X	X	X	X										X				
219					X	X		X					X		X	X		X			
221										X			X								
222													X	X	X	X	X				
224										X			X			X					
225										X			X			X	X				
228		X						X					X			X					
229										X			X		X	X					
230																					
231													X		X	X	X				
232			X	X															X		
233										X			X			X					
234													X								
235										X			X								
236			X										X			X	X				
237								X					X								
238		X	X	X	X	X	X	X					X				X			X	Uninterruptible power supplies (AC with battery backups)
239		X	X	X	X			X					X	X		X	X	X			

Table K-2. Instrumentation, Automation and Control Solids Processing (digestion, thickening, dewatering, composting, etc.) (continued).

ID	Instrumentation												Automation and Control								
Plant ID No.	Suspended Solids Analyzers	Sludge Flowmeters	Chemical Feed Flowmeters	Gas Flowmeters	Temperature Sensors	Gas Detectors	Fire Sensors	Level Sensors	Solids Volume, Weight and Concentration Meters	None	Other	Specify	Dual Power Source	Dual PLCs	Portable or In-Place Generators	Multiple Control Strategies	DCS/SCADA System Including Telemetry	Dual Field-Mounted Control Boxes	None	Other	Specify
240		X				X	X						X			X	X				
241										X					X				X		
243	X	X	X	X	X	X	X	X	X		X	Nuclear density gauges for sludge	X	X	X		X				
244		X	X	X	X			X					X			X	X				
245	X	X	X	X		X		X					X			X	X				
246										X			X			X	X				
247	X	X	X	X	X	X		X	X				X			X	X	X			
Total	**3**	**12**	**13**	**10**	**10**	**13**	**5**	**14**	**3**	**18**	**2**		**27**	**5**	**14**	**22**	**26**	**3**	**6**	**1**	

Table K-3. Reasons for Redundancy, Redundancy Ratings, and General Comments Solids Processing (digestion, thickening, dewatering, composting, etc.).

ID	Reasons for Redundancy *(Federal, State/Provincial, or Regional Regulations, etc.)*									Redundancy Ratings				Comments
Plant ID No.	Solids Processing Regulations	Discharge Permit Regulations	Safety Regulations	In-House Operations and Maintenance Requirements	IEEE Standard 493	Emergency/Disaster Planning Regulations/ Requirements	None	Other	Specify	Extremely Effective	Somewhat Effective	Currently Not Necessary	Never Used	Comments
4				X						X				Additional presses for rotation maintenance.
7				X		X								
9				X		X								
19	X	X				X				X				
21				X							X			
35				X							X			
44				X							X			
49		X		X		X				X				
60							X							
203	X		X					X	NFPA 820	X				
205		X	X	X						X				
208	X	X		X		X				X				
209				X							X			
210							X							No formal redundancy practices in place.
211	X			X							X			
212				X							X			
213				X						X				
215				X				X	Water Resources Act	X				
216							X							No redundancy
217				X						X				
218						X						X		
219	X			X						X				
221		X		X		X				X				
222			X	X		X					X			
224	X			X		X					X			

Table K-3. Reasons for Redundancy, Redundancy Ratings, and General Comments Solids Processing (digestion, thickening, dewatering, composting, etc.) (continued).

ID	Reasons for Redundancy *(Federal, State/Provincial, or Regional Regulations, etc.)*									Redundancy Ratings				Comments
Plant ID No.	Solids Processing Regulations	Discharge Permit Regulations	Safety Regulations	In-House Operations and Maintenance Requirements	IEEE Standard 493	Emergency/Disaster Planning Regulations/ Requirements	None	Other	Specify	Extremely Effective	Somewhat Effective	Currently Not Necessary	Never Used	Comments
225	X			X		X					X			
228	X	X		X							X			
229			X	X		X					X			
230														
231				X						X				
232				X							X			
233				X						X				
234								X	Good Design Practices	X				
235								X	Good Design Practices	X				
236	X			X						X				
237				X						X				
238				X						X				
239	X	X	X	X		X					X			NONE!!!
240	X	X	X	X		X				X				Additional storage space on-site.
241				X						X				Co-generation of electricity with digester gas. Boiler maintenance performed with redundant equipment.
243	X	X	X	X	X						X			
244	X	X	X							X				
245			X	X		X					X			
246	X	X		X						X				
247				X				X	State standards	X				
Total	**14**	**11**	**9**	**35**	**1**	**14**	**3**	**5**		**23**	**15**	**1**	**0**	

APPENDIX L

DISINFECTION AND EFFLUENT

Reference survey data is provided in a tabular format in this appendix for treatment facility disinfection and effluent treatment. The following tables are included:

- Table L-1. Infrastructure and Equipment
- Table L-2. Instrumentation, Automation and Control
- Table L-3. Reasons for Redundancy, Redundancy Ratings, and General Comments

Key Code:

Safety Regulations	*Occupational Health and Safety Administration—OSHA; National Fire Protection Association—NFPA; Uniform Fire Code—UFC; National Institute of Occupational Safety & Health—NIOSH*
IEEE Standard 493	Institute of Electrical and Electronic Engineers recommended practice for reliable electrical system design
Extremely Effective	Consistently use redundant systems and have been able to reduce failures with these practices
Somewhat Effective	Occasionally use redundant systems for preventative maintenance not for system failures
Currently Not Necessary	The redundant systems of this treatment process area is not currently used but may be used with future growth requirements
Never Used	The redundant systems have never been used and do not foresee a future use

Table L-1. Infrastructure and Equipment Disinfection and Effluent (chlorination, dechlorination, UV, ozone, etc.).

ID	Instrumentation														Automation and Control								
Plant ID No.	Flowmeters	Chlorine Analyzers	Dechlorination Analyzers	Ozone Analyzers	Level Sensors	Temperature Sensors	pH Analyzers	ORP Meters	Turbidity Analyzers	Gas Detectors	UV Transmittance Analyzer	None	Other	Specify	Dual Power Source	Dual PLCs	Portable or In-Place Generators	Multiple Control Strategies	DCS/SCADA System Including Telemetry	Dual Field-Mounted Control Boxes	None	Other	Specify
4			X																X				
7												X									X		
9										X					X		X	X	X				
19	X	X			X		X		X						X		X	X	X				
21																			X				
35	X	X							X						X								
44		X	X				X		X						X				X				
49	X	X					X										X		X				
60												X			X							X	UPS
203										X					X	X		X	X				
205	X	X			X		X			X					X		X	X	X				
208		X							X								X		X				
209												X					X						
210												X					X		X				
211												X						X	X			X	Take one train out of service in low flows. (winter/summer)
212												X						X					
213												X			X		X		X				
215	X	X					X											X					
216	X						X														X		
217												X									X		
218	X	X					X		X	X									X				

Water Environment Research Foundation

Table L-1. Infrastructure and Equipment Disinfection and Effluent (chlorination, dechlorination, UV, ozone, etc.) (continued).

ID	Instrumentation														Automation and Control								
Plant ID No.	Flowmeters	Chlorine Analyzers	Dechlorination Analyzers	Ozone Analyzers	Level Sensors	Temperature Sensors	pH Analyzers	ORP Meters	Turbidity Analyzers	Gas Detectors	UV Transmittance Analyzer	None	Other	Specify	Dual Power Source	Dual PLCs	Portable or In-Place Generators	Multiple Control Strategies	DCS/SCADA System Including Telemetry	Dual Field-Mounted Control Boxes	None	Other	Specify
219		X			X		X						X	dissolved oxygen analyzer	X		X	X		X			
221		X													X								
222																							
224												X			X			X					
225												X			X			X	X				
228		X	X												X			X					
229												X			X		X	X					
230																							
231																							
232	X						X			X											X		
233												X			X			X					
234												X			X								
235												X			X								
236		X	X												X		X	X	X				
237															X								
238		X	X				X								X				X			X	Uninterruptible power supplies (AC with battery backups)
239	X	X			X		X							Portable analyzers	X		X	X		X			
240										X					X			X	X				
241		X	X		X												X	X					
243	X										X				X	X	X		X				
244	X	X	X		X	X			X	X					X			X	X				

Table L-1. Infrastructure and Equipment Disinfection and Effluent (chlorination, dechlorination, UV, ozone, etc.) (continued).

ID	Instrumentation														Automation and Control								
Plant ID No.	Flowmeters	Chlorine Analyzers	Dechlorination Analyzers	Ozone Analyzers	Level Sensors	Temperature Sensors	pH Analyzers	ORP Meters	Turbidity Analyzers	Gas Detectors	UV Transmittance Analyzer	None	Other	Specify	Dual Power Source	Dual PLCs	Portable or In-Place Generators	Multiple Control Strategies	DCS/SCADA System Including Telemetry	Dual Field-Mounted Control Boxes	None	Other	Specify
245																							
246		X					X								X			X	X				
247	X	X	X		X				X						X		X	X	X	X			
Plant	12	18	8	0	7	1	12	0	7	7	1	14	1		26	2	15	20	21	3	4	3	

Table L-2. Instrumentation, Automation and Control Disinfection and Effluent (chlorination, dechlorination, UV, ozone, etc.).

ID	Instrumentation														Automation and Control								
Plant ID No.	Flowmeters	Chlorine Analyzers	Dechlorination Analyzers	Ozone Analyzers	Level Sensors	Temperature Sensors	pH Analyzers	ORP Meters	Turbidity Analyzers	Gas Detectors	UV Transmittance Analyzer	None	Other	Specify	Dual Power Source	Dual PLCs	Portable or In-Place Generators	Multiple Control Strategies	DCS/SCADA System Including Telemetry	Dual Field-Mounted Control Boxes	None	Other	Specify
4			X																X				
7												X									X		
9										X					X		X	X	X				
19	X	X			X		X		X						X		X	X	X				
21																			X				
35	X	X							X						X								
44		X	X				X		X						X				X				
49	X	X					X										X		X				
60												X			X							X	UPS
203										X					X	X		X	X				
205	X	X			X		X			X					X		X	X	X				
208		X							X								X		X				
209												X					X						
210												X					X		X				
211												X						X	X			X	Take one train out of service in low flows. (winter/summer)
212												X						X					
213												X			X		X		X				
215	X	X					X											X					
216	X						X														X		
217												X									X		

Table L-2. Instrumentation, Automation and Control Disinfection and Effluent (chlorination, dechlorination, UV, ozone, etc.) (continued).

ID	Instrumentation													Automation and Control								
Plant ID No.	Flowmeters	Chlorine Analyzers	Dechlorination Analyzers	Ozone Analyzers	Level Sensors	Temperature Sensors	pH Analyzers	ORP Meters	Turbidity Analyzers	Gas Detectors	UV Transmittance Analyzer	None	Other	Specify	Dual Power Source	Dual PLCs	Portable or In-Place Generators	Multiple Control Strategies	DCS/SCADA System Including Telemetry	Dual Field-Mounted Control Boxes	None	Other	Specify
218	X	X					X		X	X									X				
219		X			X		X						X	dissolved oxygen analyzer	X		X	X		X			
221		X													X								
222																							
224												X			X			X					
225												X			X			X	X				
228		X	X												X			X					
229												X			X		X	X					
230																							
231																							
232	X						X			X											X		
233												X			X			X					
234												X			X								
235												X			X								
236		X	X												X		X	X	X				
237															X								
238		X	X				X								X				X			X	Uninterruptible power supplies (AC with battery backups)
239	X	X			X		X							Portable analyzers	X		X	X		X			
240										X					X			X	X				
241		X	X		X												X	X					
243	X										X				X	X	X		X				

Water Environment Research Foundation

Table L-2. Instrumentation, Automation and Control Disinfection and Effluent (chlorination, dechlorination, UV, ozone, etc.) (continued).

ID	Instrumentation														Automation and Control								
Plant ID No.	Flowmeters	Chlorine Analyzers	Dechlorination Analyzers	Ozone Analyzers	Level Sensors	Temperature Sensors	pH Analyzers	ORP Meters	Turbidity Analyzers	Gas Detectors	UV Transmittance Analyzer	None	Other	Specify	Dual Power Source	Dual PLCs	Portable or In-Place Generators	Multiple Control Strategies	DCS/SCADA System Including Telemetry	Dual Field-Mounted Control Boxes	None	Other	Specify
244	X	X	X		X	X			X	X					X			X	X				
245																							
246		X					X								X			X	X				
247	X	X	X		X				X						X		X	X	X	X			
Plant	**12**	**18**	**8**	**0**	**7**	**1**	**12**	**0**	**7**	**7**	**1**	**14**	**1**		**26**	**2**	**15**	**20**	**21**	**3**	**4**	**3**	

Table L-3. Reasons for Redundancy, Redundancy Ratings, and General Comments Disinfection and Effluent (chlorination, dechlorination, UV, ozone, etc.).

ID	Reasons for Redundancy								Redundancy Ratings				
Plant ID No.	Discharge Permit Regulations	Safety Regulations	In-House Operations and Maintenance Requirements	IEEE Standard 493	Emergency/Disaster Planning Regulations/ Requirements	None	Other	Specify	Extremely Effective	Somewhat Effective	Currently Not Necessary	Never Used	Comments
4			X						X				Backup dechlorination analyzers to verify readings
7			X		X								
9			X		X				X				
19	X				X				X				
21			X							X			
35			X							X			
44			X							X			
49			X		X				X				
60			X						X				
203	X	X	X						X				
205	X	X	X						X				
208	X	X	X		X				X				
209			X							X			
210						X							No formal redundancy practices in place
211	X		X							X			
212			X							X			
213			X							X			
215			X				X	Water Resources Act	X				
216	X											X	Need to be taken out of service
217						X							
218	X				X						X		
219	X		X						X				
221	X		X						X				
222													
224	X		X		X					X			

Table L-3. Reasons for Redundancy, Redundancy Ratings, and General Comments Disinfection and Effluent (chlorination, dechlorination, UV, ozone, etc.) (continued).

ID	Reasons for Redundancy								Redundancy Ratings				
Plant ID No.	Discharge Permit Regulations	Safety Regulations	In-House Operations and Maintenance Requirements	IEEE Standard 493	Emergency/Disaster Planning Regulations/ Requirements	None	Other	Specify	Extremely Effective	Somewhat Effective	Currently Not Necessary	Never Used	Comments
225	X		X		X					X			
228	X		X		X					X			
229		X	X		X					X			
230													
231													
232	X	X	X							X			
233			X							X			
234							X	Good Design Practices					
235							X	Good Design Practices	X				
236	X								X				
237	X		X				X	Federal, state and 10-state stds. for plant power redundancy	X				
238			X						X				
239	X	X	X		X				X				NONE!!!!!!!!!!!!!!!!!!!!!!!!!!!!!!!!!
240	X	X	X		X				X				
241	X		X		X				X				Increased NPDES compliance. Dual monitoring of same paramenter has reduced false NPDES exceedances.
243	X	X	X	X						X			
244	X	X							X				
245													
246	X		X						X				
247			X				X	State standards	X				
Plant	**21**	**9**	**32**	**1**	**13**	**2**	**5**		**21**	**14**	**1**	**1**	

APPENDIX M

Regulatory Review

Contact information and reference regulatory data obtained from the U.S. state and provincial agencies are provided in this appendix.

M.1 U.S. State Regulatory Agency Contact Information

Table M-1 provides a reference guide listing of the U.S. state regulatory agencies contacted for this project. Wastewater treatment plants can reference the information provided in these tables for each U.S. state listed.

M.2 Canadian Provincial Regulatory Agency Contact Information

Table M-2 provides a reference guide listing of the Canadian regulatory agencies contacted for this project. Wastewater treatment plants can reference the information provided in these tables for each Canadian province listed.

M.3 U.S. State Regulation and Recommended Guidelines

Table M-3 provides a reference listing of the U.S. state regulations and recommended guidelines for wastewater treatment plants requiring redundancy design practices.

M.4 Canadian Provincial Regulation and Recommended Guidelines

Table M-4 provides a reference listing of the Canadian provincial regulations and recommended guidelines for wastewater treatment plants requiring redundancy design practices.

M.5 U.S. State Regional and Local Regulation and Recommended Guidelines

Table M-5 provides a reference listing of the U.S. state regional and local regulations and recommended guidelines for wastewater treatment plants requiring redundancy design practices.

M.6 Canadian Provincial Regional and Local Regulation and Recommended Guidelines

Table M-6 provides a reference listing of the Canadian provincial regional and local regulations and recommended guidelines for wastewater treatment plants requiring redundancy design practices.

Table M-1. U.S. State Regulatory Agency Contact Information.

State/Province	Agency Name/Address	Contact Person/Phone/Fax	Email/Website
Alabama	*Alabama Department of Environmental Management* 1400 Coliseum Blvd Montgomery AL 36110	*Truman Green or Glenda Dean* **Phone:** (334) 271-7700 **Fax:** (334) 271-7950	**Email:** tgg@adem.state.al.us or GLD@adem.state.al.us or Permitsmail@adem.state.al.us **Web:** http://www.adem.state.al.us/
Alaska	*Alaska Department of Environmental Conservation, Southcentral Alaska, Drinking Water and Wastewater Programs* Anchorage Office, ADEC, 555 Cordova Anchorage AK 99501	*David Johnson* **Phone:** (907) 269-7500 **Fax:** (907) 269-7600	**Email:** David_Johnson@envircon.state.ak.us or website@envircon.state.ak.us **Web:** www.state.ak.us/dec/
Arizona	*Arizona Department of Environmental Quality, Water Quality Division* 3033 N Central Ave Phoenix AZ 85012-2774	*Greg Brown, P.E., Wastewater Design Review Unit Manager* **Phone:** 602-207-4697 or (602) 207-2330 **Fax:** 602-207-4674 or (602) 207-2218	**Email:** brown.gregory@ev.state.az.us or rolfe.staci@ev.state.az.us **Web:** http://www.adeq.state.az.us/environ/water/index.html
Arkansas	*State of Arkansas Department of Environmental Quality* #1 State Police Plaza Drive Little Rock AR 72209	**Phone:** (501) 682-0656 **Fax:** (501) 682-0910	**Email:** shafii@adeq.state.ar.us or ccb@adeq.state.ar.us **Web:** http://www.adeq.state.ar.us/water/default.htm
California	*California State Water Resources Control Board, Division of Clean Water Programs* 1001 I Street Sacramento CA 95814	*Wayne Pierson* **Phone:** (916) 341-5250 or 916-341-5755 **Fax:** (916) 341-5252	**Email:** piersonw@cwp.swrcb.ca.gov or webmaster@calepa.ca.gov **Web:** http://www.swrcb.ca.gov/contact/index.html
Colorado	*Colorado Department of Public Health & Environment, Water Quality Control Division, WQCD-B2* 4300 Cherry Creek Drive South Denver CO 80246-1530	*Ron Falco* **Phone:** 303-692-3500 **Fax:** 303-782-0390	**Email:** dwrweb@state.co.us **Web:** http://water.state.co.us/

Table M-1. U.S. State Regulatory Agency Contact Information (continued).

State/Province	Agency Name/Address	Contact Person/Phone/Fax	Email/Website
Connecticut	*Connecticut Department of Environmental Protection, Bureau of Water Management* 79 Elm Street Hartford CT 06106-5127	*Roy Fredricksen* **Phone:** (860) 424-3704 **Fax:** (860) 424-4067	**Email:** roy.fredricksen@po.state.ct.us or dep.webmaster@po.state.ct.us **Web:** http://dep.state.ct.us/index.htm
Delaware	*Delaware Department of Natural Resources and Environmental Control, Division of Water Resources, Surface Water Discharges Section, Discharge Permits Program* 89 Kings Hwy Dover DE 19901	*John Defriece, P.E.,* Program Manager **Phone:** 302-739-5731 **Fax:** 302-739-8369	**Email:** John.DeFriece@state.de.us **Web:** http://www.dnrec.state.de.us/dnrec2000/ WaterResources.asp
District of Columbia	*Government of the District of Columbia, Department of Health* 5000 Overlook Avenue, SW Washington DC 20032	*James Collier* **Phone:** 202-787-2000 **Fax:** 202-787-2210	**Email:** jacollier@dchealth.com **Web:** http://www.dchealth.com
Florida	*Florida Department of Environmental Protection, Office of Ombudsman, Domestic Wastewater - Tallahassee* 3900 Commonwealth Blvd. M.S. 49 Tallahassee FL 32399	*Aaron Van Smith* **Phone:** 850-922-5485 **Fax:** 850-487-3267	**Email:** Aaron.VanSmith@dep.state.fl.us **Web:** http://www.dep.state.fl.us/default.htm
Georgia	*Georgia Environmental Protection Division* 2 Martin Luther King Jr. Drive, Suite 1152 East Tower Atlanta GA 30334	*Randy Durham* **Phone:** 404.657.5947 **Fax:** 404.651.5778	**Email:** randy_durham@dnr.state.ga.us **Web:** http://www.ganet.org/dnr/environ/
Hawaii	*State of Hawaii, Department of Health, Wastewater Branch* 919 Ala Moana Blvd, Suite 309 Honolulu HI 96814	*Dennis Tulang* **Phone:** 808-586-4294 **Fax:** 808-586-4370	**Email:** dtulang@eha.health.state.hi.us or webmail@health.state.hi.us **Web:** http://www.state.hi.us/health/eh/eiemww00.htm
Idaho	*State of Idaho, Department of Environmental Quality, Wastewater Program* 1410 N. Hilton Boise ID 83706	*Rick Huddlestun* **Phone:** (208) 373-0502 **Fax:** (208) 373-0417	**Email:** rhuddles@deq.state.id.us **Web:** http://www2.state.id.us/deq/water/water1.htm

Table M-1. U.S. State Regulatory Agency Contact Information (continued).

State/Province	Agency Name/Address	Contact Person/Phone/Fax	Email/Website
Illinois	*Illinois Environmental Protection Agency, Permit Section, Division of Water Pollution Control* 1021 North Grand Avenue East Springfield IL 62794-9276	*Chuck Fellman, P.E.* **Phone:** 217-782-0610 or 217-782-9780	**Email:** EPA1230@epa.state.il.us or Joan.Muraro@epa.state.il.us **Web:** http://www.epa.state.il.us/water/index-wpc.html
Indiana	*Indiana Department of Environmental Management, Office of Water Quality* P.O. Box 6015 Indianapolis IN 46206-6015	*Ken Lee* **Phone:** (317) 232-8660	**Email:** klee@dem.state.in.us or tmethod@dem.state.in.us **Web:** http://www.state.in.us/idem/owm/index.html
Iowa	*Iowa Water Quality Bureau Department of Natural Resources* Henry A. Wallace Bldg., 502 E. 9th Street Des Moines IA 50319-0034	*Wayne Farrand* **Phone:** 515/281-8877 **Fax:** 515/281-8895	**Email:** wayne.farrand@dnr.state.ia.us **Web:** http://www.state.ia.us/government/dnr/organiza/epd/wtrq/wtrqbur.htm
Kansas	*Kansas Department of Health and Environment, Division of Environment, Bureau of Water* 1000 SW Jackson St., Suite 420 Topeka KS 66612-1367	*K. Muelden* **Phone:** (785) 296-5500 **Fax:** (785) 296-5509	**Email:** kmuelden@kdhe.state.ks.us **Web:** http://www.kdhe.state.ks.us/water/
Kentucky	*Kentucky Division of Water, Facilities Construction Branch* 14 Reilly Road Frankfort KY 40601	*Bill Gatewood, Manager* **Phone:** 502-564-2225, ext 440	**Email:** bill.gatewood@mail.state.ky.us **Web:** http://water.nr.state.ky.us/dow/dwhome.htm
Louisiana	*Louisiana Department of Environmental Quality, Office of Environmental Services, Permits Division* 7290 Bluebonnet Blvd Baton Rouge LA 70810	*Melonie Ward, Environmental Scientist* **Phone:** (225)765-0634 **Fax:** (225)765-2725	**Email:** melonie_w@ldeq.org or webmaster@deq.state.la.us **Web:** http://www.deq.state.la.us/
Maine	*Maine Department of Environmental Protection* 17 State House Station Augusta ME 04333-0017	*Don Albert* **Phone:** 207-287-7767	**Email:** don.j.albert@state.me.us **Web:** http://www.state.me.us/dep/blwq/

Table M-1. U.S. State Regulatory Agency Contact Information (continued).

State/Province	Agency Name/Address	Contact Person/Phone/Fax	Email/Website
Maryland	*Maryland Department of the Environment* 2500 Broening Highway Baltimore MD 21224	*Dr. Ta-Shon Yu* **Phone:** 410-631-3758 or 410-631-3000	**Email:** tyu@mde.state.md.us **Web:** http://www.mde.state.md.us/environment/
Massachusetts	*Massachuettes Department of Environmental Protection Bureau of Resource Protection* One Winter Street Boston MA 02108	*Mike Rotondi, Director, Municipal Services* **Phone:** 617-292-5500	**Email:** Mike.Rotondi@state.ma.us **Web:** http://www.state.ma.us/dep/brp/brphome.htm
Michigan	*Michigan Department of Environmental Quality, Surface Water Quality Division* 120 West Chapin Cadillac MI 49601	*Dave Porter* **Phone:** 231-775-3960, ext 6261 **Fax:** 231-775-1511	**Email:** porterdr@michigan.gov or deq-ead-env-assist@michigan.gov **Web:** http://www.michigan.gov/deq/1,1607,7-135-3313---,00.html
Minnesota	*Minnesota Pollution Control Agency, Regional Environmental Management Division* 520 Lafayette Road St. Paul MN 55155-4194	*David Sahli* **Phone:** 651-296-8722 or 800-657-3864 or 651-296-6300	**Email:** david.sahli@pca.state.mn.us **Web:** http://www.pca.state.mn.us/water/index.html
Mississippi	*Mississippi Department of Environmental Quality Office of Pollution Control* 2380 Highway 80 West Jackson MS 39204	*Glen Odom or Joe Wilson* **Phone:** (601) 961-5171 **Fax:** (601) 354-6612	**Email:** Glenn_Odom@deq.state.ms.us or Joe_Wilson@deq.state.ms.us **Web:** http://www.deq.state.ms.us/newweb/opchome.nsf/pages/opc
Missouri	*Missouri Department of Natural Resources, Water Issues* PO Box 176 Jefferson City MO 65102	*Randy Clarkson* **Phone:** 573-751-3443	**Email:** nrclarr@mail.dnr.state.mo.us or oac@mail.dnr.state.mo.us **Web:** http://www.dnr.state.mo.us/water.htm
Montana	*Montana Department of Environmental Quality Community Services Bureau, Permitting and Compliance Division* 1520 East Sixth Ave Helena MT 59620	*Jim Melstad* **Phone:** 406-444-5315	**Email:** jmelstad@state.mt.us **Web:** http://www.deq.state.mt.us/index.asp

Table M-1. U.S. State Regulatory Agency Contact Information (continued).

State/Province	Agency Name/Address	Contact Person/Phone/Fax	Email/Website
Nebraska	*Nebraska Department of Environmental Quality* 1200 "N" Street, Suite 400 Lincoln NE 68509	*Charles Duerschner, Water Quality Technical Assistance or Richard Webster, Public Information Office* **Phone:** 402-471-2186 **Fax:** 402-471-2909	**Email:** MoreInfo@NDEQ.state.NE.US **Web:** http://www.deq.state.ne.us/
Nevada	*State of Nevada Division of Environmental Protection, Bureau of Water Quality Planning* 333 West Nye Lane, Suite 138 Carson City NV 89706-0851	*Joe Maez* **Phone:** 775-687-4670 **Fax:** 775-687-5856	**Email:** jmaez@ndep.state.nv.us **Web:** http://ndep.state.nv.us/bwqp/bwqp01.htm
New Hampshire	*New Hampshire Department of Environmental Services, Water Division* 6 Hazen Drive - PO Box 95 Concord NH 03302-0095	*Stephen H. Roberts, P.E., Senior Sanitary Engineer* **Phone:** 603-271-2980 or 603-271-3503	**Email:** pip@des.state.nh.us or sroberts@des.state.nh.us **Web:** http://www.des.state.nh.us/water_intro.htm
New Jersey	*New Jersey Department of Environmental Protection, Division of Water Quality* 401 East State Street Trenton NJ 08625-0029	*Chris Hoffman or Gautam Patel* **Phone:** 609-984-6840 or 609-292-4543 **Fax:** 609-984-7938	**Email:** askdep@dep.state.nj.us or Chris.Hoffman@dep.state.nj.us and Gautam.Patel@dep.state.nj.us **Web:** http://www.state.nj.us/dep/dwq/
New Mexico	*New Mexico Environment Department, Surface Water Qualtiy* Harold S. Runnels Building, 1190 St. Frances Drive Santa Fe NM 87502-0110	*Richard Rose or Gary King, Director, Facility Operations* **Phone:** 505-827-2855	**Email:** richard_rose@nmenv.state.nm.us or gary_king@nmenv.state.nm.us **Web:** http://www.nmenv.state.nm.us/swqb/swqb.html
New York	*New York State, Department of Environmental Conservation,* Division of Water 625 Broadway Albany NY 12233-3500	*William Schaff or Philip DeGaetano, Acting Director, Division of Water* **Phone:** 518-402-8116 or 518-402-8233 **Fax:** 518-402-9029	**Email:** whschaf@gw.dec.state.ny.us or dowinfo@gw.dec.state.ny.us **Web:** http://www.dec.state.ny.us/website/dow/index.html

Table M-1. U.S. State Regulatory Agency Contact Information (continued).

State/Province	Agency Name/Address	Contact Person/Phone/Fax	Email/Website
North Carolina	*North Carolina Department of Environment and Natural Resources, Division of Water Quality* 1617 Mail Service Center Raleigh NC 27699-1617	*Bill Reid or Coleen Sullins, Section Chief* **Phone:** 919-733-7015 **Fax:** 919-733-9919	**Email:** bill.reid@ncmail.net or coleen.sullins@ncmail.net **Web:** http://h2o.enr.state.nc.us/
North Dakota	*North Dakota Department of Environmental Health, Division of Water Quality* 2301 8th Ave North Fargo ND 58102	*Jeff Hauge or Gary Bracht, Wastewater Facility/Permits* **Phone:** 701-476-4121 **Fax:** 701-241-8109	**Email:** jhauge@state.nd.us or gbracht@state.nd.us **Web:** http://www.health.state.nd.us/ndhd/environ/wq/index.htm
Ohio	*Ohio EPA, Division of Surface Water,* Permit to Install and Agriculture Unit Lazarus Government Center, 122 South Front Street Columbus OH 43215	*Cathy Alexander, Supervisor or Dan Dudley, Standards and Technical Support Section 614-644-2876* **Phone:** 614-644-2021 or 614-644-2001 **Fax:** 614-644-2745	**Email:** cathy.alexander@epa.state.oh.us or dan.dudley@epa.state.oh.us **Web:** http://www.epa.state.oh.us/dsw/
Oklahoma	*Oklahoma Department of Environmental Quality, Water Quality Division* 707 N. Robinson Oklahoma City OK 73101-1677	*Dr. H. J. Thung, P.E., Engineering Manager, DWSRF / Construction Permit Section or Ed Dirhberg, Wastewater Discharge Permits* **Phone:** 405-702-8100 or 405-702-1000 **Fax:** 405-702-8101	**Email:** edward.dihrberg@deq.state.ok.us **Web:** http://www.deq.state.ok.us/WQDnew/index.htm
Oregon	*Oregon Department of Environmental Quality* 811 SW Sixth Avenue Portland OR 97204-1390	*David S. Mann* **Phone:** 503-229-5696 **Fax:** 503-229-6124	**Email:** deq.info@deq.state.or.us or MANN.David@deq.state.or.us **Web:** http://www.deq.state.or.us/about/contact.htm
Pennsylvania	*Pennsylvania DEP, Bureau of Water Supply and Wastewater Management* PO Box 8467 Harrisburg PA 17105-8467	*Ratilal Patel or Parimal Parikh* **Phone:** 717-772-4018	**Email:** rpatel@state.pa.us or pparikh@state.pa.us **Web:** http://www.dep.state.pa.us/dep/deputate/watermgt/WC/wc.htm

Table M-1. U.S. State Regulatory Agency Contact Information (continued).

State/Province	Agency Name/Address	Contact Person/Phone/Fax	Email/Website
Rhode Island	*Rhode Island, Department of Environmental Management, Office of Water Resources, Wastewater Treatment Facilities Program* 235 Promenade Street Providence RI 02908	*Alicia Good, P.E., Assistant Director or Art Zeman, P.E., Principal Engineer* **Phone:** 401-222-3961 or 401-222-4700, x7251 **Fax:** 401-222-3564	**Email:** agood@dem.state.ri.us or azeman@dem.state.ri.us **Web:** http://www.state.ri.us/dem/programs/benviron/water/index.htm
South Carolina	*South Carolina Department of Health and Environmental Control, Bureau of Water* 2600 Bull Street Columbia SC 29201	*Mike Montebello, Domestic Wastewater Permitting (803) 898-4228* **Phone:** 803-898-4300 **Fax:** 803-898-4215	**Email:** montebmj@dhec.state.sc.us **Web:** http://www.scdhec.net/water/
South Dakota	*South Dakota, Department of Environment & Natural Resources* 523 E. Capitol, Joe Foss Building Pierre SD 57501	*Kim Smith, Information Specialist or Steven M. Pirner, P.E., Secretary* **Phone:** 605-773-3152 or 605-773-3151 **Fax:** 605-773-6035	**Email:** Kim.Smith@state.sd.us or DENRINTERNET@state.sd.us **Web:** http://www.state.sd.us/denr/denr.html
Tennessee	*Tennessee Department of Environment and Conservation, Division of Water Pollution Control* 401 Church Street, 6th Floor, L&C Annex Nashville TN 37243-1534	*Roger D. Lemasters, P.E, Chief Engineer or Paul E. Davis, Director* **Phone:** 615-532-0649 or 615-532-0625	**Email:** Roger.Lemasters@state.tn.us or paul.davis@state.tn.us **Web:** http://www.state.tn.us/environment/wpc/index.html
Texas	*Texas Natural Resource Conservation Commission, Division of Water Quality* P.O. Box 13087 (MC 205) Austin TX 78711-3087	*Louis Herrin or Jim Muse, Director Policy and Regulations Division* **Phone:** 512/239-6779	**Email:** lherrin@tnrcc.state.tx.us or policy@tnrcc.state.tx.us **Web:** http://www.tnrcc.state.tx.us/water/quality/

Table M-1. U.S. State Regulatory Agency Contact Information (continued).

State/Province	Agency Name/Address	Contact Person/Phone/Fax	Email/Website
Utah	*Utah Department of Environmental Quality, Division of Water Quality* PO Box 144870 Salt Lake City UT 84114-4870	*Kiran L. Bhayani or Jay Pitkin, P.E.; Manager, Engineering and Water Quality Management Branch* **Phone:** 801-538-6080 or 801-538-6146 **Fax:** 801-538-6016	**Email:** kbhayani@deq.state.ut.us or jpitkin@deq.state.ut.us **Web:** http://www.deq.state.ut.us/EQWQ/dwq_home.ssi
Vermont	*Vermont Department of Environmental Conservation* 103 South Main Street Waterbury VT 05671-0411	*Dennis Bryer or Winslow Ladue, Policy and Planning Coordinator* **Phone:** 802-241-3735 **Fax:** 802-241-2596	**Email:** Winslow.Ladue@state.vt.us **Web:** http://www.anr.state.vt.us/dec/dec.htm
Virginia	*Virginia DEQ, Water Quality Program* 629 East Main Street Richmond VA 23219	*Cal Sawyer, P.E., Director or Cindy Berndt, Regulatory Affairs* **Phone:** (804) 698-4378	**Email:** csawyer@vdh.state.va.us or cmberndt@deq.state.va.us **Web:** http://www.deq.state.va.us/water/
Washington	*Washington Department of Ecology,* Water Quality Program 300 Desmond Drive Lacey WA 98503	*Megan White, PE, Program Manager (Lacey) or Dave Nunnallee, Municipal Unit Supervisor, Northwest Regional Office, Washington State Dept of Ecology* **Phone:** (360) 407-6405	**Email:** mwhi461@ecy.wa.gov or dnun461@ecy.wa.gov **Web:** http://www.ecy.wa.gov/programs/wq/links/program.html
West Virginia	*West Virginia DEP,* Office of Water Resources 1201 Greenbrier Street Charleston WV 25311-1088	*Allyn Turner, Director Division of Water Resources or John Morgan, Bob Coontz or Matthew Sweeney* **Phone:** 304-558-4086 or (304) 558-2107 or 304-558-0633 or 304-558-0647 **Fax:** (304) 558-5905	**Email:** aturner@dep.state.wv.us or msweeney@mail.dep.state.wv.us or bcoontz@mail.dep.state.wv.us or jmorgan@mail.dep.state.wv.us **Web:** http://www.dep.state.wv.us/item.dep?ssid=11

Table M-1. U.S. State Regulatory Agency Contact Information (continued).

State/Province	Agency Name/Address	Contact Person/Phone/Fax	Email/Website
Wisconsin	*Wisconsin Department of Natural Resources, Watershed Management, Wastewater Systems Plan Review Program* 101 S. Webster Street Madison WI 53703	*Tom A. Gilbert, Program Manager* **Phone:** 608-266-2621 **Fax:** 608-261-4380	**Email:** GilbeT@mail01.dnr.state.wi.us or behmk@dnr.state.wi.us **Web:** http://www.dnr.state.wi.us/org/water/wm/ww/
Wyoming	*Wyoming Department of Environmental Quality, Water Quality Division, Water and Wastewater Section* 122 West 25th Street, Herschler Building, 4th Floor West Cheyenne WY 82002	*Larry Robinson, Manager W&WW Div, Gary Beach, Administrator, WQ Div* **Phone:** 307-777-7075 **Fax:** (307) 777-5973	**Email:** lrobin@state.wy.us, gbeach@missc.state.wy.us **Web:** http://deq.state.wy.us/wqd/w&ww.htm

Table M-2. Canadian Provincial Regulatory Agency Contact Information.

State/Province	Agency Name/Address	Contact Person/Phone/Fax	Email/Website
Alberta	*Alberta Environment, Standards and Guidelines Branch, Alberta Environmental Protection* 5th Floor, 9820 - 106th Street Edmonton AB T5K 2J6	*Karu Chinniah* **Phone:** 780-427-8536 **Fax:** 780-422-5120	**Email:** karu.chinniah@gov.ab.ca **Web:** http://www3.gov.ab.ca/env
British Columbia	*Ministry of Water, Land and Air Protection, Province of British Columbia* 3rd Floor, 2975 Jutland Road Victoria BC V8T 5J9	*Chris Jenkins, Local Government Relations Officer* **Phone:** 250-387-6663 **Fax:** 250-387-8897	**Email:** Chris.Jenkins@gems7.gov.bc.ca **Web:** http://www.gov.bc.ca/wlap
Manitoba	*Manitoba Conservation, Environmental Approvals Branch* 123 Main Street, Suite 160 Winnipeg MB R3C 1A5	*Mike Van Den Bosch, P.Eng, Manager, Municipal, Industrial and Hazardous Waste Approvals* **Phone:** 204-945-7015 **Fax:** 204-945-5229	**Email:** mvandenbos@gov.mb.ca **Web:** http://www.gov.mb.ca/conservation/envapprovals
New Brunswick	*New Brunswick Envronment and Local Government* 20 McGloin Street Fredericton NB E3A 5T8	*Andre Chenard, Project Manager, Municipal Drinking Water Implementation* **Phone:** 506-453-3849 **Fax:** 506-453-2390	**Email:** Andre.Chenard@gnb.ca **Web:** http://www.qp.gov.bc.ca
Newfoundland	*Newfoundland Department of Environment, Water Resources Division* Sir Richard Squires Building Corner Brook NF A2H 6J8	*Ervin McCurdy, CET, Community Water and Wastewater Specialist* **Phone:** 709-637-2481 **Fax:** 709-637-2541	**Email:** emccurdy@nfld.net **Web:** http://www.gov.nf.ca/env/
Nova Scotia	*Nova Scotia Department of Environment and Labour* 5th Floor, 5151 Terminal Road Halifax NS B3J 2T8	*John Eisnor, Manager, Water and Wastewater Management Branch* **Phone:** 902-424-2282 **Fax:** 902-424-0503	**Email:** eisnorjd@gov.ns.ca **Web:** http://www.gov.ns.ca/enla/

Table M-2. Canadian Provincial Regulatory Agency Contact Information (continued).

State/Province	Agency Name/Address	Contact Person/Phone/Fax	Email/Website
Ontario	*Ontario Ministry of the Environment and Energy, Standards Development Branch* 125 Resources Road, Room N-219 Toronto ON M9P 3V6	*Dr. R. (Mano) Manoharan, P.Eng., Wastewater Treatment Specialist, Water Standards Section* **Phone:** 416-327-8128 **Fax:** 416-327-9091	**Email:** manohama@ene.gov.on.ca **Web:** http://www.gov.on.ca
Prince Edward Island	*Prince Edward Island, Fisheries, Aquiculture and Environment* Jones Building, 4th Floor, 11 Kent Street, PO Box 2000 Charlottetown PE C1A 7N8	*Morley Foy, Engineering & Utilities Section Head, Water Resources Division* **Phone:** 902-368-5036 **Fax:** 902-368-5830	**Email:** mmfoy@gov.pe.ca **Web:** http://www.gov.pe.ca/fae/
Québec	*Québec Ministère de l'Environnement, Service de l'assainissement des eaux* 675, boul. René Lévesque Est, boite 71 Québec QB G1R 5V7	*Francis Flynn, Chef de service par intérim* **Phone:** 418-521-3950 # 4989 **Fax:** 418-646-5974	**Email:** francis.flynn@menv.gouv.qc.ca **Web:** http://www.menv.gouc.qc.ca
Saskatchewan	*Saskatchewan Environment, Environmental Protection Branch* 224 - 3211 Albert Street Regina SK S4S 5W6	*Gord Will, P. Eng.* **Phone:** 306-787-6137 **Fax:** 306-787-0197	**Email:** Gwill@serm.gov.sk.ca **Web:** http://www.serm.gov.sk.ca/environment/

Table M-3. U.S. States' Redundancy Regulations and Recommended Guidelines.

State	Regulation Name	How To Access Regulation	Recommended Guidelines/Standards Name	How To Access Recommended Guidelines	Enforcement Policy
Alabama	ADEM Administrative Code R. 335-6-14-.16 (1)(l) and R. 335-6-6-.12(m)3.(i)(II). ADEM Administrative Rule 335-6-6-.12(e) and 335-6-5-.15(5) (Proper Operation and Maintenance for both NPDES and SID permitted facilities)	ADEM Administrative Rules are posted on the ADEM web page at www.adem.state.al.us	There are no guidelines or procedures separate from the regulations.		Only required under Clean Water State Revolving Loan (CWSRF) program review and in defense of a bypass in NPDES boilerplate language based on regulations. The regulations are administered through the NPDES and SID permits programs.
Alaska	No specific regulatory requirement in State Wastewater Disposal Regulations (18 AAC 72) for redundant components. However 18 AAC 72.220 requires that treatment works design conform to sanitary engineering principles and practice.		"Recommended Standards for Sewage Works" Great Lakes-Upper Mississippi River Board of State Public Health and Environmental Managers (GLUMRB); "Wastewater Engineering: Treatment, Disposal, Reuse" Metcalf and Eddy "Wastewater Treatment Plant Design, Manual of Practice Number 8" WPCF	Commonly held references or they can be purchased.	All plans for treatment works must be approved by State Wastewater engineers (Alaska Department of Environmental Conservation) prior to construction.
Arizona	Arizona Administrative Code R18-9-b202(A)(10) Yes - Requirement for a Standby power source	By Web: http://www.sosaz.com/public_services/Title_18/18-09.htm			Through Permit Process - Aquifer Protection Permit
Arkansas			10-State Standards adopted by ADPCE	Get a copy through www.hes.org or call 518-439-7286	Issuing a construction permit

Table M-3. U.S. States' Redundancy Regulations and Recommended Guidelines (continued).

State	Regulation Name	How To Access Regulation	Recommended Guidelines/Standards Name	How To Access Recommended Guidelines	Enforcement Policy
California		The Regulations concerning water reclamation can be found at http://www.dhs.ca.gov/ps/ddwem/publications/waterrecycling/purplebookupdate6-01.PDF, and the regulation concerning surface impoundments can be found at http://www.calregs.com/cgi-bin/om_isapi.dll?clientID=1144323907&infobase=ccr&softpage=Browse_Frame_Pg42			
Colorado			Water Quality Control Commission Policy 96-1, Design Criteria for wastewater treatment plants	Website http://www.cdphe.state.co.us/op/wqcc/wepol.htm	They are not enforceable but we essentially have authority to disapprove of designs that are not in conformance with the criteria.
Connecticut			TR-16, "Guide for the Design of Wastewater Treatment Works" by the NEIWPCC		Plans and specification review of all projects.

Table M-3. U.S. States' Redundancy Regulations and Recommended Guidelines (continued).

State	Regulation Name	How To Access Regulation	Recommended Guidelines/Standards Name	How To Access Recommended Guidelines	Enforcement Policy
Delaware					State law (De. Code, Title 7, Part 7, Chapter 60, Section 6003, available at www.michie.com) and regulations do require a State permit to construct wastewater collection, transmission, or treatment facilities. Our group issues those permits, considering "10-State Standards" and "good engineering practice" (GEP). On a case by case basis, the State can require redundant systems in the construction permit approval process. Current and proposed De. State Regulations Governing Control of Water Pollution Control, respectively, are on-line at http://www.dnrec.state.de.us/water2000/Sections/SurfWater/Library/RGCWP.pdf and http://www.state.de.us/research/register/july2000/DNREC_-_Water_Pollution.PDF. Regarding the "Ten-State Standards" (from http://www.state.ia.us/epd/wastewtr/tenstate.htm), "A copy may be obtained by contacting the Health Education Services at Albany, New York. Cost per copy Telephone number is 518-439-7286.
District of Columbia			10-State Standards		

Table M-3. U.S. States' Redundancy Regulations and Recommended Guidelines (continued).

State	Regulation Name	How To Access Regulation	Recommended Guidelines/Standards Name	How To Access Recommended Guidelines	Enforcement Policy
Florida	Rule 62-600, Florida Administrative Code (F.A.C.), adopts EPA's "Design Criteria for Mechanical, Electric, and Fluid System and Component Reliability" manual. Rule 62-600, F.A.C., (http://fac.dos.state.fl.us/) also requires that all facilities for which the Department received complete permit applications after July 2, 1991 provide a minimum of Class III reliability. The rule specifically requires Class I reliability for certain types of wastewater treatment facilities such as those providing reclaimed water for Part III public access reuse systems, disposing effluent through under-ground injection wells, and discharging to Class I waters. Wastewater treatment plants discharging to other surface waters are required to provide either Class I or II reliability in accordance with EPA's "Design Criteria for Mechanical, Electric and Fluid System and Component Reliability." Public access reuse systems include irrigation of areas accessible to the public, irrigation of edible crops, fire protection, aesthetic purposes (such as fountains or decorative ponds) or other activities. The Department can approve alternate methods of providing Class I, II, and III reliability if the permittee provides reasonable assurances that the level of reliability provided is equivalent to the class of reliability required.	Rule 62-600, F.A.C., can be found on the web at http://fac.dos.state.fl.us/. Copies of the "Design Criteria for Mechanical, Electric, and Fluid System and Component Reliability" manual are available at our offices and we provide copies if requested.			A DEP permit is required to build a new facility or expand/modify an existing facility. The Department ensures reliability requirements are met during the permitting process.
Georgia			EPA's Design Criteria for Mechanical, Electric, and Fluid System and Component Reliability	Located on the Web and Georgia EPD makes copies for the cities or engineers that want it.	We perform individual design reviews on new facilities and modifications

Table M-3. U.S. States' Redundancy Regulations and Recommended Guidelines (continued).

State	Regulation Name	How To Access Regulation	Recommended Guidelines/Standards Name	How To Access Recommended Guidelines	Enforcement Policy
Hawaii	There are a few redundancy requirements such as a 100% back up for ground disposal systems [section 11-62-25 (a)(1)] and reference to the City and County of Honolulu design standards [section 11-62-24 (b)].	The website for the state rules is: www.state.hi.us/doh/rules/11-62.pdf			All design plans for wastewater treatment systems must be reviewed and approved by our agency. All newly constructed wastewater treatment systems must be inspected and approved for use by our agency.
Idaho	No Response Received.				
Illinois	The regulation is at 35ILC (Illinois Administrative Code) Section 306.102(a)	The regulations can be obtained from the State of Illinois or they are on the web at http://www.epa.state.il.us. Click on "Rules and Regulations" then click on "Illinois Environmental Regulations" (the top one on the list).	Our design standards elaborate on where stand-by power or pumping capability is required. These standards are at 35 ILC Section 370.450 for pumping stations and at 35 ILC Section 370.550(a) for treatment works. I should explain that our design standards are not guidelines or recommendations: they were adopted by the agency and they are regulations.		It is included in our review procedures for issuance of state permits that are required for the construction or modification of pumping stations and treatment facilities.
Indiana			Recommended Standards for Wastewater Facilities, 1997 Edition	Copies can be ordered from Health Education Services PO Box 7126 Albany, NY 12224 Phone: 518-439-7286 Web: www.hes.org	Handled during the review of plans and specifications when issuing state construction permits for wastewater treatment facilities.

Water Environment Research Foundation

Table M-3. U.S. States' Redundancy Regulations and Recommended Guidelines (continued).

State	Regulation Name	How To Access Regulation	Recommended Guidelines/Standards Name	How To Access Recommended Guidelines	Enforcement Policy
Iowa	Rule 567 IAC 64; Redundancy (reliability) criteria are found throughout the standards but are mainly discussed in Chapter 14 for treatment plants and in Chapter 13 for pumping stations. See specifically Sections 14.5, 13.4.1, and 13.11.	Rules can be accessed by links through the Department's Environmental Protection Division website at: www.state.ia.us./epd The link for the rule is www.legis.state.ia.us/Rules/2002/iac/567iac/56764/56764.pdf The link for the Iowa Wastewater Design Standards is www.state.ia.us/epd/wastewtr/design.htm	The Iowa Wastewater Design Standards do not include criteria specifically for the SBR process. The state does have design guidance for the process.	www.state.ia.us/epd/wastewtr/sbr0301.pdf	Construction permits are required for construction, installation or modification of any wastewater disposal system or part thereof. Generally, existing facilities not meeting a current design standard are not required to unless/until a project is undertaken. It is possible that inconsistent or noncompliant treatment could be the impetus for a state directive for a facility upgrade or construction which must then meet current standards.
Kansas			Kansas has published documents which are referenced by state regulations. The documents are called "Minimum Standards of Design" and discusses a variety of design features including redundancy. The Minimum Standards generally describe "good or accepted engineering practices which are accepted by the regulatory agency. However, the Minimum Standards recognize there are circumstances where alternate designs or new products impact design. There is a Minimum Standard published for Wastewater Systems (Municipal and Industrial), Livestock Waste Systems, and Public Water Supplies.	Hard copy	Enforcement is generally through permitting procedures, i.e., to receive a permit the plans and specifications must generally comply with Minimum Standards. Usually compliance would be achieved by modifying the design to address the deficiency, or put another way, to show compliance with Minimum Standards. Another approach could be to issue the appropriate permit with a condition addressing the issue.
Kentucky	Kentucky Regulation 401KAR:005, Sections 13 & 20	http://www.lrc.state.ky.us/regulations/401/005/005.htm			Through review of plans and specifications during the issuance of construction permits.

Table M-3. U.S. States' Redundancy Regulations and Recommended Guidelines (continued).

State	Regulation Name	How To Access Regulation	Recommended Guidelines/Standards Name	How To Access Recommended Guidelines	Enforcement Policy
Louisiana	Our regulations address redundant treatment systems in only a vague manner. In LAC 33:IX.2355.E under Proper Operation and Maintenance the regs says, "The permittee shall at all times properly operate and maintain all facilities and systems of treatment and control (and related appurtenances) which are installed or used by the permittee to achieve compliance with conditions of this permit. Proper operation and maintenance also includes adequate laboratory controls and appropriate quality assurance procedures. This provision requires the operation of back-up or auxiliary facilities or similar systems which are installed by a permittee only when the operation is necessary to achieve compliance with the conditions of the permit." The bypass provisions of our regs also touch on redundant treatment equipment. LAC 33:IX.2355.M.4.a says, "Bypass is prohibited, and the state administrative authority may take enforcement action against a permittee for bypass, unless: (i) bypass was unavoidable to prevent loss of life, personal injury, or severe property damage; (ii) there were no feasible alternatives to the bypass, such as the use of auxiliary treatment facilities, retention of untreated wastes, or maintenance during normal periods of equipment downtime. This condition is not satisfied if adequate back-up equipment should have been installed in the exercise of reasonable engineering judgment ....."				
Maine			TR-16 Guides for the Design of Wastewater Treatment Works, 1998	Contact the New England Interstate Environmental Training Center (NEIETC) at 2 Fort Road, South Portland, Maine 04106	Section 412(b) provides that we review the design of wastewater treatment works. We use TR-16 as guidelines.

Table M-3. U.S. States' Redundancy Regulations and Recommended Guidelines (continued).

State	Regulation Name	How To Access Regulation	Recommended Guidelines/Standards Name	How To Access Recommended Guidelines	Enforcement Policy
Maryland			Design Guidelines for Sewerage Facilities and Design Guidelines for Waste-water Pumping Stations for the Protection of Shellfish Waters and Swimming Waters. (These guidelines provide definitive instructions of what redundant facilities are required.)	Engineering consultants who design wastewater treatment facilities or wastewater pumping stations have copies of design guidelines sited. For specific details the interested party can contact Dr. Ta-Shon Yu at 410-631-3758.	There are 4 steps to implement redundancy: 1. Pre-design conference; 2. Issuance of a permit for construction; 3. Construction inspection; 4. Certification of the constructi completion as the permit issuec
Massa-chusetts			We do have the draft 1988 "Guidelines For The Design, Construction, Operation & Maintenance Of Small Sewage Treatment Facilities With Land Disposal" that includes redun-dancy requirements, which is available by hard copy, but it is outdated and is currently being revised. I am working on a new draft now which I hope to have out to a select internal group by the end of the month. I don't expect it to be final and/ or available at least until the end of the year, maybe longer.	Hard copy	We check redundancy by reviewing the plans and specifications. If we see a CWMP, then we make sure that redundancy is addressed in the preliminary design.
Michigan			Recommended Standards for Wastewater Facilities (10-State Standards)	Order from Health Education Services, A Division of Health Research, Inc., PO Box 7126 Albany, NY 12224 Phone: 518-439-7286 Web: www.hes.org	We use case by case review of facilities. Part 41 gives the state broad powers to review the operation and design of waste-wtaer treatment facilities and we may 'order' facilities to provide appropriate redundancy.
Minnesota			10-State Standards; MPCA Reliability Guidelines; and other design references such as WEF MOP8, and Metcalf & Eddy Wastewater Engineering	http://www.pca.state.mn.us/water/pubs/rmwtp.html	Through the State's permitting and design approval processees. No construction can begin until we issue the permit and design approval.

Table M-3. U.S. States' Redundancy Regulations and Recommended Guidelines (continued).

State	Regulation Name	How To Access Regulation	Recommended Guidelines/Standards Name	How To Access Recommended Guidelines	Enforcement Policy
Mississippi			The State of Mississippi's Department of Environmental Quality has "Guidance" for the construction of domestic wastewater treatment systems. There is no specific reference to redundancy. There are areas where two or more units are recommended, but no reference that each unit be capable of handling the total design flow, except for collection system pumping stations where two pumps are minimum, each able to handle the design flow.	A copy of our "Guidance" can be received from Rickey Terry, Chief of the Municipal and Private Facilities Branch at 601-961-5115, and Rickey_Terry@ deq.state.ms.us	Most rely on our review of the systems, which is required by regulation.
Missouri			We have design rules that require redundancy consistent with 10-States Standards. These are in Rules of the Department of Natural Resources, Division 20 — Clean Water Commission, Chapter 8—Design Guides.	They are available on the Secretary of States website.	Missouri Clean Water Law RsMO 644 requires construction permits for all wastewater activities. The review process includes checking to see if redundancy consistent with the rules is provided.
Montana			Montana Department of Environmental Quality Circular DEQ 2 Design Standards for Wastewater Facilities	Via Website at: http://www.deq.state.mt.us/wqinfo/Circulars/DEQ2.PDF	Submittal of Engineering Reports, plans, specifications and designs.

Table M-3. U.S. States' Redundancy Regulations and Recommended Guidelines (continued).

State	Regulation Name	How To Access Regulation	Recommended Guidelines/Standards Name	How To Access Recommended Guidelines	Enforcement Policy
Nebraska	Nebraska's rules and regulations for the design of wastewater works are found in Title 123, Rules and Regulations for Design, Operation and Maintenance of Wastewater Treatment Works. Title 123 has no direct requirements for redundancy in designing wastewater facilities.		Title 123 states that the Ten-States Standards are recommended as a guidance document in designing wastewater works. There are several redundancy requirements in the 10-States Standards. Nebraska maintains a set of design guidelines for various wastewater treatment systems. NDEQ's main interest in redundancy, as you define it, is in providing standby power for pumping and treatment units so that sewage backups or bypasses can be avoided.	The public may request a written copy of any NDEQ guideline be mailed to them at no cost. They will be placed in the Departments web page in the near future.	NDEQ has construction permit authority for all wastewater works designed and constructed in the state. Designs that are inadequate are brought up to minimum standards when the plans and specifications are reviewed.
Nevada	NAC 445a.284 list that the state will use 10-States Standards as a review guideline. These standards list redundancy requirements for pumping and aeration systems.		10-States Standards		As part of the required state plan review for all treatment work, NDEP's engineers review all work using its standard references. Additionally, the state has several desgin guidances which it has produced to cover the design of various treatment systems. These are available at the following: www.ndep.state.nv.us. and are listed in the Bureau of Water Pollution Control.

Table M-3. U.S. States' Redundancy Regulations and Recommended Guidelines (continued).

State	Regulation Name	How To Access Regulation	Recommended Guidelines/Standards Name	How To Access Recommended Guidelines	Enforcement Policy
New Hampshire	NH Department of Environmental Services' (NHDES') technical design standards (including redundancy requirements) for wastewater treatment facilities are codified as NH Administrative Rules at Chapter Env-Ws 700, entitled "Standards of Design and Construction for Sewerage and Wastewater Treatment Facilities".	These and other NHDES rules are available for viewing/downloading at the NHDES web site (www.des.state.nh.us), and are also available for hard-copy purchase at NHDES' Public Information Center (Tel. 603-271-8876).			Under (variously) NH state laws, NHDES administrative rules and requirements of NHDES/EPA grant/loan programs, engineering designs for any proposed sew-erage (including WWTFs) must be submitted to NHDES for review/ approval prior to installation. NHDES engineering staff examine the design submittals for compli-ance with (redundancy and other design) requirements of afore-mentioned Chapter Env-Ws 700.
New Jersey	No, except for UV disinfection	www.state.nj.us/dep/ dwq/twa	We require designs to handle at least 250% to design flow in multiple units.	www.state.nj.us/dep/dwq/twa	Treatment Works Approval (TWA) Program.
New Mexico			Our guidelines ("Guidelines for Wastewater Collection and Treatment Works in New Mexico) have no specific redundancy requirements. The guideline was issued in 1987 and we are now in the process of updating them using the current version of the 10-State Standards as a starting point. I hope to have the revised guidelines out for public comment this fall.	Hard copies are available upon request. The new revised guidelines will posted on our website.	Publicly funded wastewater project plans and specifications are reviewed by the Construction Programs Bureau and comments could include redundancy issues if deemed appropriate.
New York			New York has adopted the design guidance contained in the 1997 Edition of the Recommended Standards for Wastewater Facilities (10-States Standards). This document contains redundancy requirements.	The 10-States Standards document is available on the New York State web site at http://www.dec.state.ny.us under division of water/ wastewater permits/ referneces.	The Environmental Conservation Law requires that all wastewater treatment facilities be approved by the Department. Design reports and construction documents are reviewed to ensure conformanc with our design criteria which include redundancy requirements.

Table M-3. U.S. States' Redundancy Regulations and Recommended Guidelines (continued).

State	Regulation Name	How To Access Regulation	Recommended Guidelines/Standards Name	How To Access Recommended Guidelines	Enforcement Policy
North Carolina	15A NCAC 2H .0200 - Waste Not Discharged to Surface Waters (Section .0219 - Minimum Design Requirements), and 15A NCAC 2H .0100 - Wastewater Discharges to Surface Waters (Section .0124 - Reliability)	Hard copies of these rules have been widely available for years. Copies may also be accessed from our web site at: http://h2o.enr.state .nc.us/admin/rules/	N/A - There are regulations addressing treat/equipment redundancy in NC.		Through review and approval of wastewater treatment plant design plans prior to construction and inspection of operating facilities.
North Dakota					N/A
Ohio			Great Lakes - Upper Mississippi River Board of State Public Health Officials and Environmental Managers - Recommended Standards for Wastewater Facilities, 1997 (aka 10 State Standards)	Ohio State offers to provide copies of pertinent pages when we can. The 10-States Standards can be ordered from Health Education Services in Albany, New York.	Ohio requires approval of plans prior to construction, and during the review process we ensure that the recommended redundancy is included (or there is adequate justification for not following the recommendations).
Oklahoma	OAC Chapter 656	Website: www.deq.state.ok.us Under rules/regulations and chapter 656			Based on compliance evaluation inspection, field inspection, perforamnce failures, plant improvement/modification and change of design standards.

Table M-3. U.S. States' Redundancy Regulations and Recommended Guidelines (continued).

State	Regulation Name	How To Access Regulation	Recommended Guidelines/Standards Name	How To Access Recommended Guidelines	Enforcement Policy
Oregon	We rely on EPA 1974 regs, EPA-430-99-74-001, and current practice. We have not had any professional objections to them or problems with the 1974 regs.	Very difficult for them. Their consulting engineers normally have it. On occasion we may make a copy for an engineer of EPA-430-99-74-001, or refer them to NTIS to get their own.	We do have state standards for sewage pump stations, not treatment plants. We reviewed the Washington State standards in 1998, and generally concur with them so far as they go. However, since we have state engineers that review all treatment plant modifications, EPA's 1974 redundancy/reliability standards are imposed fairly uniformly statewide.		State law (statute) requires our agency to approve all sewage works prior to construction. Oregon admin rules implementing this law are at www.deq.state.or.us/wq/wqrules/340Div52.pdf. Oregon DEQ employs a few experienced, licensed engineers to review all municipal sewage plant plans, including modifications and upgrades. Formal approval letters are issued for each project. Note these regs are not applied to industrial plants or similar non-residential establishments where the place can simply be shut during upsets and failures, or for maintenance, repair, or upgrade of its treatment works.
Pennsylvania			Domestic Wastewater Facilities Manual (ID 362-0300-001)	Department's Website at www.dep.state.pa.us or by contacting DEP to obtain a hard copy	The Department has a permitting program in place. A Water Quality Management permit (Part II) is required for the construction and operation of any treatment facility. An inspection of the facility is made before it is placed in use.

Table M-3. U.S. States' Redundancy Regulations and Recommended Guidelines (continued).

State	Regulation Name	How To Access Regulation	Recommended Guidelines/Standards Name	How To Access Recommended Guidelines	Enforcement Policy
Rhode Island			Rhode Island has no regulations for Redundancy Design Practice Specifications, although the State's Water Quality Regulations (Rule 15) indicate that our review utilizes Guides for the Design of Municipal Treatment Works (TR-16). TR-16 emphasizes the need to design for "the greatest operating convenience, flexibility, and economy"	TR-16 can be obtained from the New England Interstate Environmental Training Center 2 Fort Road Portland, Maine 04106. Their tel. # is 207-767-2539. In addition Rhode Island's Water Quality Regulations may be downloaded from our website: www.state.ri.us/dem	When reviewing any wastewater treatment or conveyance system, our office looks for redundant units such that forward flow or treatment is possible with one unit out of operation. If the applicant does not conform to our requirements, then we will not issue an "Order of Approval." The Order of Approval is basically a write-off from our office that we have reviewed the project technically and find it in conformance with our guidelines and best engineering practices. The Order of Approval allows the applicant to proceed to construction. If an applicant proceeds to construction without the Approval, then they are in violation of the Water Quality Regs.
South Carolina	R61-67.400 Standards for Wastewater Facility Construction. (Note the title of section 400 is Relability Classifications) Applies to all new WWTF's and expanding facilities.	Website See http://www.scdhec.net/water/html/laws.html			The regulations apply to engineering design and construction of all wastewater treatment facilities. The implemetation is via a Construction Permit that is required for all new, expanding or upgrading WWTFs. There is a separate Enforcement Section to address compliance with existing NPDES & construction regulations. There are also District offices (12 Districts in SC) that must issue an Approval to Operate (after Construction is completed and the project inspected) for Construction Permits before they may proceed.

Table M-3. U.S. States' Redundancy Regulations and Recommended Guidelines (continued).

State	Regulation Name	How To Access Regulation	Recommended Guidelines/Standards Name	How To Access Recommended Guidelines	Enforcement Policy
South Dakota			Recommended Design Criteria for Wastewater Collection and Treatment Facilities (1991)	access- www.state.sd.us/denr/des/P&s/ designcriteria/designT.html	State Statute 34A requires submittal of plans and specifications for SDDENR review and approval.
Tennessee			There are design criteria in place. The publication is entitled "Design Criteria for Sewage Works".	The Division of Water Pollution Control is in the process of placing these criteria on our website, a process that will take some additional time. Alternatively, the "Design Criteria" can be distributed in electronic format via e-mail or on a 3.5" disc. (in response to a request).	The "Design Criteria" are intentionally written as a non-regulatory document. However, it is written to indicate best design practices. We expect consulting engineers to design wastewater treatment facilities in accordance with the provisions in the "Design Criteria". However, we also recognize that wastewater treatment technology is constantly evolving. We have the expertise in-house to evaluate any design that deviates from our "Design Criteria". If that deviation appears to be technically sound and capable to complying with permit requirements and best engineering practices, we can approve technology that is either not included in or appears to violate parts of our "Design Criteria". If that deviation does not appear to be appropriate, we will not approve it for construction. There is a regulation that specifically requires all modifications to sewage systems to be approved by the State before construction begins. As such, consultant engineers know to contact us first before construction actually begins.
Texas	Chapter 317 - Design Criteria for Sewerage Systems, Texas Natural Resource Conservation Commission	Regulations can be accessed from the TNRCC agency website http://www.tnrcc.state.tx.us			Permits, Design Engineer Reports, etc.

Table M-3. U.S. States' Redundancy Regulations and Recommended Guidelines (continued).

State	Regulation Name	How To Access Regulation	Recommended Guidelines/Standards Name	How To Access Recommended Guidelines	Enforcement Policy
Utah	Utah State Rule R317-3 Design Requirements	Via web www.deq.state.ut.us or you can go to www.utah.gov or http://www.deq.state.ut.us/EQWQ/documents/DOC_RULE.HTM or call me Kiran L. Bhayani, PE, DEE, Manager, Design Evaluation. POB 144870, Salt Lake City, Utah 84114-4870, at 801 538 6080, or email at kbhayani@utah.gov	R317-3 various sections cover it, refer to treatment works for specific directions.	via web http://www.deq.state.ut.us/EQWQ/documents/DOC_RULE.HTM or call me Kiran L. Bhayani, PE, DEE, Manager, Design Evaluation. POB 144870, Salt Lake City, Utah 84114-4870, at 801 538 6080, or email at kbhayani@utah.gov	O&M Inspections, Review of operating reports, discharge monitroing reports, and lastly; through composite correction programs and enforcement actions if necessary and inevitable.
Vermont			Basis for Final Design	U.S. Mail	During reviews of designs for upgrade/improvement projects.
Virginia	Virginia Sewage Collection and Treatment (SCAT) Regulations implemented by the Virginia Department of Health (VDH), Division of Wastewater Engineering (DWE).	http://vdhWEB/oehs/onsite/Regulations/SCAT-Final.pdf			See supplemental information in Appendicies
Washington	However, our state plan review regulation, WAC 173-240-040 (2), makes direct reference to our design criteria (discussed below) to assure that engineering plans and reports "are reasonably consistent" with them.	WAC 173-240 can be viewed at: http://www.ecy.wa.gov/biblio/wac173240.html	The redundancy requirements are based on EPA's "Reliability Classification", which is repeated almost verbatim in our state design criteria manual, titled "Criteria for Sewage Works Design". The criteria and the resultant redundancy requirements are found in section G2-8 of our design criteria.	This document can be viewed on-line at: http://www.ecy.wa.gov/biblio/9837.html This document will open in Acrobat Reader. Jump to page 114 of 482 to go directly to the section on reliability classification and redundancy requirements.	The design criteria are used whenever engineering reports, or plans & specs, are reviewed for a sewage works project. Normally, the redundancy/reliability issue is not triggered until a treatment facility expands or upgrades, thus requiring a formal review & approval process as mandated in Washington state law RCW 90.48.110.
West Virginia	No Response Received.				

Table M-3. U.S. States' Redundancy Regulations and Recommended Guidelines (continued).

State	Regulation Name	How To Access Regulation	Recommended Guidelines/Standards Name	How To Access Recommended Guidelines	Enforcement Policy
Wisconsin	Chapter NR 110, Wisconsin Administrative Code. Also s. NR 210.08, W.A.C.	Wisconsin's statutes and codes can be accessed from a state website at http://www.legis.state.wi.us/rsb/code			Implementation is primarily through NR 110 plan reviews conducted for any proposed new or modified sewerage system project. If permit violations are occurring, improved emergency operating provisions may be required as part of WPDES permit conditions or as part of enforcement actions.
Wyoming	Sections 11, 23 and 30, Chapter 11 Wyoming Water Quality Rules and Regulations.	Website: http://soswy.state.wy.us/cgi-win/SSGI_2.exe			Construction Permits

Table M-4. Canadian Province Redundancy Regulations and Recommended Guidelines.

State/Province	Regulation Name	How To Access Regulation	Recommended Guidelines/ Standards Name	How To Access Recommended Guidelines	Enforcement Policy
Alberta	Alberta Wastewater and Storm Drainage Regulation states that all wastewater systems are designed and operated in accordance with the requirements stipulated in the document titled Standards and Guidelines for Municipal Waterworks, Wastewater, and Storm Drainage Systems.	Website: http://www3.gov.ab.ca/env/info/infocentre/publist.cfm search under "wastewater municipal" for the document.			
British Columbia	BC has taken a performance based approach using their Waste Management Act to enforce the Municipal Sewage Regulation (MSR) which addresses design and construction of facilities. In the BC MSR, redundancy is referred to as reliability and is covered under Schedule 7 Section2(6) and Appendix 1.	The MSR is on-line at http://www.qp.gov.bc.ca/statreg/reg/W/WasteMgmt/129_99.htm			Plant designs are submitted and when accepted, a certificate of approval issued for the plant.
Manitoba	Authority exists that could be used to require redundancy if required but there is no provincial regulation that mandates redundancy.		There are no guidelines. Our practice is to required redundancy when good engineering practice is not followed. Examples: - If a lift station is designed with only one pump and there is a risk of sewage overflow or backup that could cause public health or environmental problems, we would require redundancy. - If components of a treatment process could fail during winter conditions and are not enclosed within protective structures to allow for repairs or repairs cannot be carried out by other means, we would consider requiring multiple units to provide for some redundancy.		As previously stated, redundancy can be required through our environmental review and licensing process as required. Our belief is that good engineering practice should address most redundancy requirement before it a design is submitted for environmental approval.

Table M-4. Canadian Province Redundancy Regulations and Recommended Guidelines (continued).

State/Province	Regulation Name	How To Access Regulation	Recommended Guidelines/ Standards Name	How To Access Recommended Guidelines	Enforcement Policy
New Brunswick			Parts of our guidelines deal with Redundancy Design Practices when we deal with the construction of sewage pumping stations and certain components such as mechanical bar screens, mechanical grit removal, certain parts of the aeration equipment, etc.	We are presently working with two sets of Guidelines: "Guidelines for the Collection and Treatment of Wastewater, 1987" and the "Atlantic Canada Standards and Guidelines Manual". Both are available through our department.	Regulation 82-126 under the Clean Environment Act requires the proponent to have an approval to construct before any construction of a wastewater collection or treatment facility is undertaken. Conditions set in this document are binding. During the review of the proposal should the project not adhere to the conditions set in the guidelines this is corrected by adding this as a condition of the approval to construct.
Newfoundland			Guidelines for the Design, Construction and Operation of Water and Sewerage Systems, Government of Newfoundland, Department of Environment, Section 5: 4.3.3 Multiple Units: "Multiple Units capable of independent operation are mandatory. The use of single unit treatment works shall only be considered if other provisions have been included to assure continuity and flexibility of treatment. The Guidelines for the Design, Construction and Operations of Water and Sewerage Systems are being reviewed/revised, and will include updated requirements. The design review process prior to issuing a permit to construct addresses this concern. As well, the Permit to Operate also addresses the issue.	Through the Department of Environment, during review for permitting. Most consultants involved in design of municipal water and wastewater works have a copy of the guidelines.	Regulation 82-126 unnder the Clean Environment Act requires the proponent to have an approval to construct before any construction of a wastewater collection or treatment facility is undertaken. Conditions set in this document are binding. During the review of the proposal should the project not adhere to the conditions set in the guidelines this is corrected by adding this as a condition of the approval to construct.

Table M-4. Canadian Province Redundancy Regulations and Recommended Guidelines (continued).

State/Province	Regulation Name	How To Access Regulation	Recommended Guidelines/ Standards Name	How To Access Recommended Guidelines	Enforcement Policy
Nova Scotia			Nova Scotia Standards and Guidelines Manual for the Collection Treatment and Disposal of Sanitary Sewage (1992).	This document can be purchased from the province's on-line bookstore (www.gov.ns.ca)	Wastewater treatment plant designs must go through a provincial approval process before the plant is built to ensure the design meets the terms of reference of the standards and guidelines manual.
Ontario			Internal Standards Development Branch policy documents.	Contact the Standards Development Branch and request a copy.	The design, and thus, the redundancy design practices, is subjected to a formal review procedure and a certificate of approval is issued for the treatment plant.
Prince Edward Island			PEI presently uses the "Atlantic Canada Standards and Guidelines Manual for Collection, Treatment and Disposal of Sanitary Sewage".	Contact Environment Canada, Atlantic Region office in Halifax, NS. The contact is Paul Klaamas who can be reached at paul.klaamas@ec.gc.ca	The plant design is submitted by the municipality and the redundancy design practices approved when the certificate of approval is issued for the treatment plant.
Québec	We do not have redundancy design practices in regulation for wastewater treatment facilities. In the pulp and paper industry we ask the mills to have emergency basin or lagoon.		We do not have specific guide-lines, etc., but in the process of authorization of the project, if the engineer has not planned to install some equipment in double, we ask him to do it, specially for pumps, blowers, and other mechanical equipment. Most of the project that is present to authorization have redundancy in major mechanical and other equipment. We think that is good engineering practice.		If it's necessary, the recommended practices should be used.

Table M-4. Canadian Province Redundancy Regulations and Recommended Guidelines (continued).

State/Province	Regulation Name	How To Access Regulation	Recommended Guidelines/ Standards Name	How To Access Recommended Guidelines	Enforcement Policy
Saskatchewan	Redundancy issues are addressed in design guidelines and through approval to construct process.	Publication information is available by contacting the department.	The design guidelines are sited as "A Guide to Sewage Works Design".	Publication information is available by contacting the department.	Based on requirement to obtain approval to construct and revisited during permit renewal and inspections. The department has general authority to require specific actions be taken if need to protect public health or the environment.

Table M-5. U.S. States' Regional and Local Redundancy Regulations and Recommended Guidelines.

State/Province	Regional/Local Regulation Name	How To Access Regulation	Regional/Local Recommended Guidelines/ Standards Name	How To Access Recommended Guidelines	General Comments
Alabama	In the case of SID permits, redundancy may be required by local pretreatment authority.				
Alaska	There are none. The State of Alaska directly implements these regulation on a statewide basis.				
Arizona					
Arkansas					
California					
Colorado					
Connecticut			TR-16 is a regional publication generally in use by most New England States. Each New England State had a representative on the document review board.		
Delaware	Public Health regs. require one day storage capacity for wastewater treatment plants that discharge to "Harvestable Shellfish Waters".				I believe the U.S. EPA (www.epa.gov) has nume publications with recommendations re. wastewater treatment plant design and operation.
District of Columbia					
Florida					
Georgia					Not sure about Local Regional information. The reason redundancy was implemented in State is from the grant days of the 70s.

Table M-5. U.S. States' Regional and Local Redundancy Regulations and Recommended Guidelines (continued).

State/Province	Regional/Local Regulation Name	How To Access Regulation	Regional/Local Recommended Guidelines/ Standards Name	How To Access Recommended Guidelines	General Comments
Hawaii	The city standards are not on the website. You can call James Honke at (808) 5275040 or write to Tim Steinberger at City and County of Honolulu, 1000 Uluohia St., Suite 308, Kapplei, HI 96707				
Idaho					No Response Received.
Illinois					I think a few comments are in order to clarify my answers. The operative words in your definition are "...units that are not used during normal average operations." Under that definition, the only units for which we have redundancy requirements are: stand-by generators or stand-by engine driven pumps at pumping stations and some treatment plants; and a second connection to a public utility source at pumping stations and some treatment plants. We do require that pumping stations be able to pump the maximum design flow with the largest unit out of service. In that case, however, the pumps are alternated so that all the pumps are used during normal operations. The same goes for blowers at activated sludge plants.
Indiana					
Iowa					
Kansas			There are the "10-State Standards" which are commonly referenced. These are not rules or regs and even though they are called "Standards". I would describe them as accepted or recommended practices.		
Kentucky					For questions 6 & 7, None that we know of.
Louisiana					

Table M-5. U.S. States' Regional and Local Redundancy Regulations and Recommended Guidelines (continued).

State/Province	Regional/Local Regulation Name	How To Access Regulation	Regional/Local Recommended Guidelines/ Standards Name	How To Access Recommended Guidelines	General Comments
Maine			TR-16	Contact the New England Interstate Environmental Training Center (NEIETC) at 2 Fort Road, South Portland, Maine 04106	
Maryland	Most counties and municipalities have design guidelines of their own. They are consistent with the State's guidelines.			It does not matter whether the regional or local entities have redundancy guidelines or not. Maryland Department of the Environment controls the redundancy requirements through the legal process of issuing a construction permit with implementation procedures.	
Massachusetts	For redundancy as well as general design guidelines, we use TR-16: Guides for the Design of Wastewater treatment Works published by New England Interstate Water Pollution Control Commission as is available for purchase as hard copy. We have also referred to an old EPA publication: Design Criteria For Mechanical, Electric and Fluid System and Component Reliability (EPA-430-99-74-001). If it's still in print, EPA would have it.	Hard copy			We do not have any state regulations dealing with redundancy or other design practices.

Table M-5. U.S. States' Regional and Local Redundancy Regulations and Recommended Guidelines (continued).

State/Province	Regional/Local Regulation Name	How To Access Regulation	Regional/Local Recommended Guidelines/ Standards Name	How To Access Recommended Guidelines	General Comments
Michigan					Michigan is probably a different case than most states. We are a member of the "Great Lakes - Upper Mississippi River Board of State and Provincial Public Health and Environmental Managers" who produce a document called RECOMMENDED STANDARDS FOR WASTEWATER FACILITIES. Our administrative rules state that the department shall consider design criteria as set forth in RECOMMENDED STANDARDS FOR SEWAGE WORKS (now RECOMMENDED STANDARDS FOR WASTEWATER FACILITIES) – See following rule: PART 41 (Parts 1 to 5). SEWER-AGE SYSTEMS (Promulgated pursuant to Part 41, Sewerage Systems, of the Natural Resources and Environmental Protection Act, 1994 PA 451, as amended) R 299.2936 Plans and specifications; minimum requirements. Rule 36. Information con-tained within the plans and specifications submitted to the department for review and approval pursuant to R 299.2933 shall include, but not necessarily be limited to, the material recommended in section 12, plans, and section 13, specifications, of the recommended standards for sewage works R 299.2938 Plans and specifications; review by department. Rule 38. (1) Upon receipt of plans and specifications for the con-struction or alteration of a sewerage system or portion thereof, the department shall review them as soon as practicable to determine their completeness with regard to the minimum requirements specified in R 299.2936 and their acceptability with regard t accepted design standards for wastewater facilities in this state. In making its review, the department shall consider design criteria as set forth in recommended standards for sewage works and shall be assured that the sewerage system or portion thereof is so designed so as to protect the public health and prevent unlawful pollution. We also have review checklists that address redundancy for the various wastewater system components.

Table M-5. U.S. States' Regional and Local Redundancy Regulations and Recommended Guidelines (continued).

State/Province	Regional/Local Regulation Name	How To Access Regulation	Regional/Local Recommended Guidelines/ Standards Name	How To Access Recommended Guidelines	General Comments
Minnesota					In Minnesota we do not have rule regarding technical design for wastewater treatment. Our primary recommendations come from 10-States Standar and we have supplemented those with some of our own more specific design criteria. We do have MPCA system reliability guidelines that address redundancy that are on our website.
Mississippi	We know of no jurisdiction in Mississippi that has any redundancy requirements.				
Missouri	There are no redundancy requirements of any sort other then our design rule.				
Montana					
Nebraska	We are not aware of regional or local authorities in Nebraska that have redundancy requirements, as you define the term.		We are not aware of regional or local guidelines for redundancy.		
Nevada	Many of Nevada's Counties do have design requirements.				
New Hampshire	While not utilized in NH, other New England states reportedly employ Technical Report #16 (TR-16) design standards as developed by the New England Interstate Water Pollution Control Commission (255 Ballardvale Street, Wilmington, MA 01887). DES does refer to this and other references (such as "10-State Standards" and notable engineering texts) for interim guidance in the absence of specific adopted standards.				

Table M-5. U.S. States' Regional and Local Redundancy Regulations and Recommended Guidelines (continued).

State/Province	Regional/Local Regulation Name	How To Access Regulation	Regional/Local Recommended Guidelines/ Standards Name	How To Access Recommended Guidelines	General Comments
New Jersey					
New Mexico					
New York	We are unaware of any local redundancy regulations.		We are unaware of any local redundancy guidelines.		
North Carolina	No, there are none. The State rules cited above apply to industrial and municipal wastewater treatment plants, which I understand are the focus of this study.				
North Dakota					N/A
Ohio					Some counties and cities do have approval requirements for systems within their boundaries, and I would expect that some of them have their own redundancy requirements. However, I am not familiar with their individual requirements and can not speak positively on that issue. If you need more information on this, let me know and I can put you in contact with some of our district staff (or the city/county staff) who could tell you for sure.
Oklahoma			Recommended Standards for Wastewater Facilities	www.hes.org	

Table M-5. U.S. States' Regional and Local Redundancy Regulations and Recommended Guidelines (continued).

State/Province	Regional/Local Regulation Name	How To Access Regulation	Regional/Local Recommended Guidelines/ Standards Name	How To Access Recommended Guidelines	General Comments
Oregon	No DEQ's design requirements and review practices are applied uniformly statewide. There are no regional jurisdictions with intervening responsibility for review. Note that we have exempted 90% final review responsibility for a few of the larger cities that have asked for such an exemption and qualify for it. That is, they have adequate engineering staff and agree to stringent conditions in exchange for exemption from final review. Only the final review is delegated, control over facilities design plans and specifications is never delegated. Thus in Oregon, municipal plan review exemptions do not delegate responsibility for reliability, redundancy, effectiveness, operability, maintainability, flexibility, durability of materials, or any other aspect of the design. DEQ believes the state law assigning the review/ approval process to the agency does not authorize it to delegate this responsibility to permittees.	The gritty details of this topic are found in www.deq.state.or.us/wq/wqrules/340Div52.pdf, specifically in OAR 340-52-045(2) and (3).			
Pennsylvania			Wastewater treatment facilities are required to comply with state regulations and we are not aware of any regional or local regulations concerning this issue.		

Table M-5. U.S. States' Regional and Local Redundancy Regulations and Recommended Guidelines (continued).

State/Province	Regional/Local Regulation Name	How To Access Regulation	Regional/Local Recommended Guidelines/ Standards Name	How To Access Recommended Guidelines	General Comments
Rhode Island	No, there are no local or regional redundancy regulations.		TR16, referred to in response #1, was compiled by officials and engineers from all the New England States, plus New York. I am not certain if the other states utilize TR-16 as their design guidelines or regulations exclusively or at all. You will need to check with our counterparts in those states.		
South Carolina	None at the State level specifically for WWTFs although there may be requirements in other program areas for activities like Chemical Storage, Underground or Above Ground Storage Tanks, Incinerators (or other fuel fired systems) that may need other permits (like Air Permits, etc.)	This is the general website address for the different environmental program areas. http://www.scdhec.net/eqc/admin/html/permtype.html	You would have to query the individual Program -areas.	Individual POTWs may establish their own redundancy requirements internally but I am not aware of any other local or regional groups with guidelines.	
South Dakota					
Tennessee					
Texas					

Table M-5. U.S. States' Regional and Local Redundancy Regulations and Recommended Guidelines (continued).

State/Province	Regional/Local Regulation Name	How To Access Regulation	Regional/Local Recommended Guidelines/ Standards Name	How To Access Recommended Guidelines	General Comments
Utah	All regulations or requirements are at State level. There are district engineers in the field, however, bulk of the compliance oversight is provided from Salt Lake City office, and new construction/installation is reviewed and permitted before installation or construction by the State. Please visit our home page http://www.deq.state.ut.us for getting familiar with the organization and programs.				
Vermont					
Virginia					For questions 6 & 7 local government information for the State of Virginia can be accessed on the following website: http://www.vipnet.org/portal/government/localgov.htm
Washington	None that I am aware of.		None that I am aware of. Redundancy requirements are at the state level.		
West Virginia					No Response Received.
Wisconsin					For questions 6. & 7. - I know of no regional or l regulations for redundancy design practices in Wisconsin, but some designers (especially for larger facilities) may develop and use project specific redundancy criteria that differ or go beyond the minimum state standards. I've seen designers refer to the EPA technical bulletin: "Design Criteria for Mechanical, Electric, and Fluid System and Component Reliability" (EPA-430-99-74-001), but this is now a rather old document and I'm not sure if has ever been upd
Wyoming					

Table M-6. Canadian Regional and Local Redundancy Regulations and Recommended Guidelines.

State/Province	Regional / Local Regulation Name	How To Access Regulation	Regional/Local Recommended Guidelines/Standards Name	How To Access Recommended Guidelines	General Comments
Alberta					
British Columbia					
Manitoba			I expect that the Manitoba Water Service Board and some municipalities have design and construction specifications that address redundancy issues.	I do not have specific knowledge of these specification documents so you will have to contact these agencies or municipalities directly. With respect to treatment facilities accessing the information, these agencies and municipalities are normally the owners or design commissioning agencies therefore access is not an issue.	
New Brunswick	We are not aware of any.		We are not aware of any.		
Newfoundland	We are not aware of any.		We are not aware of any.		
Nova Scotia					
Ontario			Larger municipalites and regional governments have review procedures which would incorporate redundancy design.	Direct contact with the municipalities required to obtain guidelines.	
Prince Edward Island					
Québec					
Saskatchewan	Province is the only regulatory authority.				

APPENDIX N

Regulatory Supplemental Materials

Supplemental regulatory information and references for U.S. State and Provincial agencies are provided in this appendix.

N.1 U.S. State Regulations or Recommended Guidelines

Supplemental regulatory information is provided for the following U.S. States.

N.1.1 Minnesota

Minnesota's Pollution Control Agency provides the following checklists for various wastewater treatment processes to its sanitary districts. These checklists provide a good example of providing safeguards for implementing redundancy in the design and operations of the wastewater treatment facilities.

N.1.1.1 Minnesota's Facility Plan Submittal Checklist

Minnesota Pollution Control Agency

Facilities Plan Submittal Checklist

Submissions Required for a Complete Facilities Plan (Minn. R. 7077.0272)

Water/Wastewater/Financial Assistance/#2.02, February 2002

FACILITY INFORMATION

Project Name ____________________

Proposed Dates for Construction ____________________

City's Authorized Representative ____________________

Title ____________________ Phone __________

Mailing Address ____________________ __________

Technical Agent or Consulting Engineer ____________________

Name of Firm/Organization ____________________ Phone __________

PLEASE CIRCLE YES OR NO FOR THE FOLLOWING QUESTIONS:

Is the Facilities plan signed by an engineer registered in the State of Minnesota? **Yes** **No**

Has the municipality in which the facility will be located held at least one public hearing to discuss the proposed project?

Yes **No** **Date hearing was held** __________

CHECK THE BOXES BELOW IF YOU HAVE INCLUDED THE FOLLOWING ITEMS:

Please see Minn. R. 7077.0272 for more information about the content of facilities plan.

- ☐ A summary of the public hearing documenting that the following items were discussed:
  - ☐ The various treatment alternatives considered
  - ☐ The location of the project site
  - ☐ The reasons for choosing the selected treatment method
  - ☐ The estimated sewer service charges
- ☐ A summary of the comments received at the public hearing and the action taken to address those comments
- ☐ A complete list of addresses used for public notice purposes on a form **provided by the MPCA**
- ☐ A copy of the resolution of the municipality's governing body adopting the facilities plan
- ☐ A list of ordinances or intermunicipal agreements required for the implementation and administration of the project
- ☐ A signed treatment agreement with each significant industrial user
- ☐ A completed environmental information sheet (referred to as an EIW) **provided by the MPCA**
- ☐ For individual sewage treatment systems that serve more than one structure, an assurance from the municipality stating that all property owners who will be served by the proposed system agree to be part of the system, to participate in the construction project, and to finance future operation, maintenance, and replacement of the system
- ☐ Copies of all notifications, certifications, and comments received (required to meet the provisions of Minn. Stat. § 116.182)

If all of the above items are not included with the facilities plan, the facilities plan is incomplete and may be returned or filed until a complete submittal is received. Facilities plan review will not begin until a complete submittal is received.

wq-wwtp#2.02 Minnesota Pollution Control Agency, 520 Lafayette Road North, St. Paul, Minnesota 55155-4194
(651) 296-6300, toll-free (800) 657-3864, TTY (651) 282-5332 or (800) 657-3864

This material can be made available in alternative formats for people with disabilities.

N.1.1.2 Minnesota's Wet Well, Raw Sewage Pumps, Bar Screens, and Comminutor Review Checklist

Minnesota Pollution Control Agency

Water Quality

Wastewater Technical Review and Guidance

WET WELL, RAW SEWAGE PUMPS BAR SCREENS AND COMMINUTOR - REVIEW CHECKLIST

Water/Wastewater/#5.59, May 2001

FACILITY NAME ____________________ DATE ____________________

CONSULTING ENGINEER ____________________ SITE INSPECTION (DATE & INSPECTOR) ____________________

PLANNING OR DESIGN PHASE ____________________

Inflow Characteristics:
Average Wet Weather Flow (gpm) ____________________
Peak Hourly Flow (gpm) ____________________

Wet Well

Dimensions ____________________

Capacity (gal) ____________________

Detention Time:
@Average Wet Weather Flow ____________________
@ Peak Hourly Flow ____________________

Flow Measurement Device ____________________

Is there an adequate alarm system? ____________________

Is there emergency power? ____________________

Adequate Number Air Changes? ____________________

Control Float Tubes placed to Minimize affects of flow? ____________________

Contents:

wq-wwtp5-59

Minnesota Pollution Control Agency, 520 Lafayette Road North, St. Paul, Minnesota 55155-4194
(651) 296-6300, toll-free (800) 657-3864, TTY (651) 282-5332 or (800) 657-3864
This material can be made available in alternative formats for people with disabilities.

Printed on recycled paper containing at least 20 percent fibers from paper recycled by consumers.

Raw Sewage Pumps

Number ____________________

Capacity (each pump) ____________________

Variable Speed
Pumps? ____________________

Bar Screens

Number of Units ____________________

Width of Units ____________________

Bar Size (in.) ____________________

Bar Spacing (in.) ____________________

Flow through Velocity
(ft. per. sec.)
@ Average Wet Weather Flow ____________________
@ Peak Hourly Flow ____________________

Slope ____________________

Flow Measurement Device ____________________

Is bar screen mechanically cleaned or manually cleaned ____________________

Anticipated problems with access to screens or ease of removal of screenings? ____________________

Final disposal of screenings? ____________________

Comminutor

Number of Units ____________________

Capacity
(each unit) ____________________

Are auxiliary screens proposed for alternate use? ____________________

N.1.1.3 Minnesota's Biological Treatment Checklist for Suspended Growth Activated Sludge and Oxidation Ditches

Minnesota Pollution Control Agency

Water Quality

Wastewater Technical Review and Guidance

BIOLOGICAL TREATMENT CHECKLIST – SUSPENDED GROWTH – ACTIVATED SLUDGE, OXIDATION DITCHES

Water/Wastewater/#5.10, April, 2001

FACILITY NAME ____________________ DATE ____________________

CONSULTING ENGINEER ____________________ SITE INSPECTION (DATE & INSPECTOR) ____________________

PLANNING OR DESIGN PHASE ____________________

INTRODUCTION

The activated sludge process is an aerobic biological process, which uses the metabolic reactions of microorganisms to attain an acceptable effluent quality by removing substances exerting an oxygen demand. There are many modifications of activated sludge that can be used to meet specific treatment requirements. These modifications consist of and are defined as follows:

Conventional Activated Sludge - The conventional activated sludge process consists of an aeration tank, a secondary clarifier and a sludge recycle line. Both influent sewage and recycled sludge enter the tank at the head end, move through the tank in plug flow and are aerated for a period of time.

Step Feed Activated Sludge - The Step feed process (sometimes called Step Aeration) is a modification of the activated sludge process in which sewage is introduced at two or more points in the aeration tank while return sludge is introduced only at the head end of the aeration tank. In this modification, the oxygen demand is more uniformly spread over the length of the aeration tank.

Tapered Aeration Activated Sludge - The tapered aeration activated sludge process is a modification of the conventional activated sludge process. Tapered aeration affects only the arrangement of the aeration devices in the aeration tank. In tapered aeration the diffusers or aeration devices are spaced closer together at the head end of the tank to match the oxygen demand.

Contents:

wq-wwtp5-10

Minnesota Pollution Control Agency, 520 Lafayette Road North, St. Paul, Minnesota 55155-4194
(651) 296-6300, toll-free (800) 657-3864, TTY (651) 282-5332 or (800) 657-3864

Complete Mix Activated Sludge - The complete mix activated sludge process is characterized by distributing the influent sewage, return sludge and air equally throughout the aeration tank to form a nearly homogeneous mixture in the tank. The organic load and thus oxygen demand is uniform from one end of the tank to the other.

Contact Stabilization Activated Sludge - The contact stabilization activated sludge process utilizes two aeration compartments to divide the two phases of BOD removal. The first phase, adsorption, takes place in the first aeration tank (contact tank). Mixed liquor from the contact tank then flows to the clarifier. Return sludge flows to the second aeration tank (reaeration tank) where the second phase, absorption, occurs. The reaeration tank provides detention time before flowing into the contact tank.

Extended Aeration Activated Sludge - The extended aeration activated sludge process operates in the indigenous respiration phase of the growth curve, which necessitates a relatively low organic loading (low F/M) and long aeration time.

Oxidation Ditch Activated Sludge - The oxidation ditch activated sludge process is an extended aeration process utilizing a ring-shaped channel and aeration than provides continuous flow around the channel.

Pure Oxygen Activated Sludge - Pure oxygen activated sludge is characterized by the use of high purity oxygen instead of air for aeration. 30

I. **DESCRIPTION OF FACILITY** – Provide a short description of the proposed facility:

____
____
____

Hydraulic and Organic Loading / Design Flow

Average Wet Weather (AWW)	____________	gals/day
Peak Hourly Wet Weather (PHWW)	____________	gals/day
Peak Instantaneous Wet Weather (PIWW)	____________	gals/day
Organic Loading, CBOD5	____________	lbs/day
Explain Primary modification of activated sludge	____________	
Type of activated sludge process chosen	____________	
Type of Aeration	____________	

Tank Dimensions and Volume	____________	
Detention time in aeration basin		hrs
Aeration tank organic loading (CBOD/day/1000ft^3	____________	
Design F/M ratio	____________	
Design MLSS	____________	
Process Preceded by (specify process)	____________	
Process Followed by (specify process)	____________	

Final effluent limitation	________ mg/L CBOD5
	________ mg/L TSS
	________ mg/L Ammonia (min. concentration)
	________ mg/L Phosphorus

GENERAL

Calculations and/or documentation shall be submitted to justify the basis of design for the following:

a. Process efficiency. ________
b. Aeration tanks. ________
c. Aeration equipment (including oxygen and mixing requirements) ________
d. Secondary clarifiers ________
e. Return sludge equipment ________
f. Waste sludge equipment ________
g. Sludge handling facilities ________

Wastewater must be amenable to biological treatment. Check for industrial discharges that may cause a nutrient imbalance, variable pH or temperature, or toxic or slug discharge. Certain dairy wastes, paper wastes, potato wastes particularly should be examined for nutrient deficiencies. ________

City has sufficient equipment, trained perator, financial resources, etc. to properly operate and mange this system: ________

Efficient use of energy is provided?
Particular attention should be given to initial operation when oxygen demands may be significantly less than the design oxygen demand. The design should always maintain the minimum mixing levels; mixing may control power requirements at low oxygen demands. Energy conservation measures should be considered in design of aeration systems. For diffused aeration systems, the following should be considered: ________

a. Use of smaller compressors and more units. ________
b. Variable-speed drives on positive-displacement compressors. ________
c. Intake throttling on centrifugal compressors. ________
d. Use of high-efficiency diffusers. ________

For mechanical aeration systems, the following should be considered:

a. Use of smaller aerators. ____________
b. Variable aeration tank weirs. ____________
c. Multiple speed motors. ____________
d. Use of timers. ____________

Duplication of critical units (pumps, aerators, motors) as well as standby units (tanks etc.) have been provided. See "Reliability Guidelines." ____________

Standby or auxiliary power is provided for aeration and pumping equipment. TYPE ____________

Flexibility allows for at least three modes of operation (two for extended aeration). Flexibility refers to providing several modification, series and parallel operation, munerous return and waste sludge flow options, etc. If feasible, aeration tank systems, except extended aeration, should be designed to accommodate at least three (3) modes of operation (such as plug flow, complete mix, contact-stabilization, step feed etc.). Two (2) modes of operation for extended aeration systems larger than 1893 m3/day (0.5 MGD) should be considered. Design of aeration systems should provide adequate flexibility to vary the oxygen transfer capability and power consumption in relation to oxygen demands. MODES ____________

Protection of aeration tanks, clarifiers and equipment from freezing. Winter protection for the aeration tanks may consist of windwalls, earthen bank insulation, and extra freeboard, etc. Other equipment such as blowers, pumps, etc. should also be protected. Secondary clarifiers and aerobic digesters should be covered under certain conditions. TYPE ____________

PRETREATMENT

Grit removal, scum and comminution or screening ____________

AERATION

Number of Units – The requirements for dual aeration tanks shall be as listed in MPCA's Reliability for Mechanical Wastewater Treatment Plants Guidelines.

a. Additional aeration tanks designed specifically for nitrogen removal shall not be used for satisfying the requirement of dual aeration tanks for secondary treatment activated sludge.

b. Plants with initial flows of less than one-half design should have at least two (2) aeration basins. All flow splitting devices shall have vertical inlet into the splitter and split by overflowing horizontal surfaces at the same elevation.

c. When two (2) or more tanks are installed, both series and parallel flow schemes should be available. In addition each train of aeration tank(s) and clarifier(s) shall have its own separate return and waste sludge pumping facilities, as well as intergroup return and wasting sludge capabilities.

ARRANGEMENT OF AERATION TANKS

a. **Dimensions** – Deep aeration tanks utilizing surface aeration should be avoided without assurance of adequate mixing. Spiral flow refers to the circular flow pattern that results what the diffusers are located along one wall of the aeration tank. Fillets should be provided around the bottom of aeration tanks where the wall and bottom meet to eliminate dead spots.

Do any unusual factors exist that will affect design such as excessive diurnal load (>4:1, <2:1) variations, stricter degree of treatment than 25 mg/1 BOD, 30 mg/l TSS, temperatures, pH, etc. Wastes with high concentrations of BOD diurnal load ratios or peak hourly BOD5 to average BOD5 of greater than 4:1. may be handled best by complete mix activated sludge because BOD is dispersed completely through tank and makes maximum use of dilution. Contact Stabilization may not be practical for wastewater with a high soluble BOD such as dairy waste. Lab tests should be performed on mixtures of industrial and domestic wastes. Conventional activated sludge may be designed for slightly higher organic loadings if the system also has the capability to operate in a mode that can handle higher loadings. For step feed plants, influent hydraulic design should permit all flow to enter any single pass. The flow to each pass must be controllable. ____________________

CRITERIA

Liquid depth in aeration tank, should not be less than 10 or more than 30 feet. Horizontally mixed aeration tanks shall have a depth of not less than 5.5 (as an exception). ________ FT ________

Shape of tank and installation of aeration equipment shall avoid short-circuiting. ____________________

Access must be available to allow for observing aeration tank contents, mixing pattern, sampling, etc., in aeration tank and settling, scum collection, level weirs, etc., in clarifier. __________________

Each compartment has load-bearing walls for individual compartment dewatering. Slope bottom to assure complete drainage. Aeration tank drainage should be discharged upstream of the activated sludge process.

Piping for bypassing and dewatering each individual compartment is provided? __________________

Pressure relief valves in each compartment provided? Tanks with freeboard greater than 18' shall provide rails, steps or other means of egress from each aeration tank. Additional freeboard or windbreak may be necessary to protect against freezing or windblown spray. Walkways should be located and/or designed to preclude spray and foam from freezing on their surfaces. Locating walkways away from or sufficiently elevated, above water levels and using grated walkways are examples of methods used to prevent ice build up. __________________

Aeration Tank freeboard, minimum 18 inches. ______ IN ____________

If a mechanical surface aerator is used, the freeboard should not be less than 3 feet to protect against windblown spray freezing on walkways, etc. __________________

AERATION EQUIPMENT-DIFFUSED AIR

In diffused air systems, a minimum air flow volume of 20-30 scfm/1000 ft^3 than volume is required for adequate mixing velocities and to avoid depositing of solids. For a grid aeration system, mixing rates of 10 to 15 $ft^3/min/1000ft^3$ is sufficient. __________________

A bypass to the atmosphere should be installed avoid surging. The bypass should be arranged to be open at starting and at all times when the airflow is less than that at the surge point. __________________

Aluminum pipe should be avoided due to corrosion problems. __________________

Centrifugal blowers should and rotary positive displacement blowers must be provided with both inlet and discharge silencers to reduce the high pitched whine that is transmitted to the piping.	
Surging will result If the blower is throttled on the discharge side or by adjustment of the air valves at the aeration thanks to capacity equal to or less than that at the peak of the head-capacity curve.	
Total aeration tank volume shall be divided between two or more units, capable of independent operation, for reliability.	
Both parallel and series operation are available when two or more aeration tanks exist.	
Inlets and outlets of aeration tank shall be equipped with valves, gates, plats, weirs or other devices to control flow to any unit and to maintain constant liquid level.	
The effluent weir for a horizontally mixed aeration tank system must be easily adjustable by mechanical means and shall be sized based on the design peak instantaneous flow plus the maximum return sludge flow.	
Design peak instantaneous flow can be carried with any single aeration tank unit out of service.	
All channels and pipes designed to maintain self-cleansing velocities. Min.2'/Sec	FT/SEC
Channels not being used during alternate flow patterns can be drained	
Minimum dissolved oxygen concentration that can be maintained by aeration equipment. Should be at least 2 mg/l D.O.	Mg/l
Design oxygen requirements in lbs. O_2/lb. Peak BOD5 applied to aeration tanks (Minimum 1.1 lbs. O_2/lb. BOD_5, except extended aeration the value shall be 1.5).	LBS 0_2 DAY
For **nitrification** oxygen demand of recycle flow must be added to determine 0_2 requirement	LBS 0_2
Total oxygen requirement, BOD recycle plus **nitrification** in lbs. 0_2/lbs. BOD.	LBS 0_2 DAY

Alpha and Beta factor (50% of clean water efficiency where 90 % + waste is domestic sewage, higher industrial percentages should use less that 50 % of clean water efficiencies and calculations need be submitted to justify)	ALPHA BETA
Certified aeration device transfer efficiency.	
Air requirements for diffused air systems, ft^3/lb BOD peak aeration tank loading. (500 ft^3/lb BOD except extended aeration which shall be 2050 ft^3/lb BOD.)	CUBIC FT/LB BOD
Air requirements for additional air for channels, pumps, digesters, etc.	CUBIC FEET/MIN
Total air requirements, ft^3/Min.	CUBIC FEET/MIN
High air intake temperature for blower (should be designed for 115°F (45°C)).	F°
Low air intake temperature for motor drive (should be designed for -20°F (-29°C))	F°
Blowers in multiple units and can meet maximum air demand with single largest unit out of service	
Air rate shall be variable or adjustable. For energy conservation	
Adequate workspace shall be provided around each blower and other equipment (2' minimum). For easy maintenance access.	FT
Diffuser systems designed for diurnal peak oxygen demand or 200% of design average oxygen demand, whichever is greater.	PEAK OR 200%
Total head loss from blower outlet of silencer to diffuser inlet (should not exceed 0.5 psi at average operating conditions.	psi
Spacing of diffusers should be in accordance with oxygen requirements through length of tank and designed to facilitate adjustment of aeration throughout tank.	
Diffusers are removable without dewatering aeration tank.	
Hoist provided to raise diffusers.	

Individual diffusers are equipped with control valves with indicator markings for throttling, or for complete shutoff. ________________

Air filters with easy access provided for blowers. ________________

Soundproofing provided between blower room and other work areas. ________________

Airflow rate meters provided for each aeration compartment in addition to a total air flow rate meter. ________________

AERATION EQUIPMENT – MECHANICAL

Design Requirements – Provide mixing with use of a draft tube. Mixing requirements vary from .75 to 1.5 hp/1000ft^3 tank volume for complete mixing. Adjustable aeration tank outlet weirs are desirable with mechanical aerators for varying submergence of the aerator. Aeration devices in small plants may have time minute increments.
Reliability – An additional motor (and bearings for brush aerator systems) shall be provided for single mechanical aerator systems.

Oxygen requirements, lbs. 92/Hr. DAY ________________

Certified testing of mechanical aerator and drive unit shall verify lbs. 0_2/Hr. ________________

In the absence of specific design information, the oxygen requirements shall be calculated using a transfer rate not to exceed 2 lbs. Of oxygen per horsepower per hour in clean water under standard test conditions. ________________

Minimum D.O. maintained by mechanical aeration system (2 mg/l). mg/l ________________

Mechanical aeration system can maintain all biological solids in suspension (for a horizontally mixed aeration tank system an average velocity of 1 ft. per second must be maintained). ________________

Can mechanical aeration system meet maximum oxygen demand and maintain process performance with largest unit out of service. ________________

Provide for varying the amount of oxygen transferred (adjustable mechanism, adjustable aeration tank outlet weirs, time clocks for smaller facilities, etc.) ________________

Provide that motors, gear housing, bearings, grease fittings, etc., and be easily accessible and protected from inundation and spray as necessary for proper functioning of the unit. Aerators shall be removable without dewatering tanks on shutting off air supply to other tanks. ______

Protection from freezing and excessive ice build up. ______

RETURN SLUDGE EQUIPMENT

Return Sludge Rate – Timers should be considered only as a last resort. Timers should allow for operation in five (5) minute increments. Airlift pumps used for return sludge and waste sludge shall have air flow regulators with gauged handles for control of return sludge and waste sludge rates.
Return Sludge Pumps - Air lifts commonly plug up at top, so access must be provided at top of airlift to remove accumulated debris.

Return Sludge Piping - Observation and sampling of waste sludge, as well as return sludge, should be provided from a safe vantage point. Return sludge should be added to the influent wastewater prior to entering the aeration tank or added to the aeration tank in a manner and location that assures adequate mixing with the influent waste.

Minimum return sludge rate as percentage of design average flow (DAF). ______ %

Maximum return sludge rate (%) ______ %

Method of varying rate of sludge return (variable speed motors, drives or timers; timers for small facilities and as a last resort only) obtained with largest return pump. ______ TYPE

Can maximum return sludge capacity be obtained with largest return pump blower out of service? ______

Size of suction and discharge openings on return sludge pumps (three-inch minimum). ______ IN

Do airlifts allow rapid and easy cleaning? ______

Diameter of airlift (minimum three inches). ______ IN

Diameter of return sludge discharge (four-inch minimum). ______ IN

Design velocity of return sludge piping at normal return sludge rtes (minimum two feet per second). FPS

Devices provided for observing, sampling, measuring and controlling return sludge flow from each settling tank

OBS.
SAMP.
MEAS.
CONT. ______

Type of waste sludge handling. ____________________

Type of waste sludge control (manual, timer, % of return sludge, etc.) ____________________

Maximum capacity of waste sludge control facilities as percentage of average rate of sewage flow (not less than 25%). ______ %

Minimum capacity of waste sludge control facilities as percentage of average rate of sewage flow (0.5 % or 10 gal/min, whichever is larger). ______ %

Devices provided for observing, sampling, controlling and measuring waste activated sludge flow (list type separately).
OBS.
SAMP.
MEAS.
CONT ____________________

Where is waste sludge discharge? ____________________

MEASURING DEVICES

Measurement devices should be located to measure flows to each tank unit. This means if flow splitting is utilized, measurement should be located downstream of the flow splitting device to measure flows after splitting. When a totalizer and recorder are available to measure flows prior to flow spilling, a weir may be sufficient to satisfy flow measurement after splitting.

For facilities over 1 MGD, can measuring devices for return and waste sludge also total and record? ____________________

OXIDATION DITCHES

In addition to the above criteria concerning activated sludge facilities, oxidation ditches should conform to the following recommendation.

Sidewalls and bottoms of oxidation ditch channels are tied together with reinforcing bar or are poured as on slab to eliminate frost heave?
There must be something holding the walls and bottom together to prevent frost heaving causing separation. The weight sufficient to prevent this separation. ____________________

Type of backfill for oxidation ditch channel (should be sand with drainage).

Type of cover ground between channels.
The areas between the channels should be covered with an impermeable material to prevent rainwater from entering the ground between the channels. TYPE ____________________

Location of influent and return sludge lines with respect to rotor (lines should be upstream for best mixing). ______________________

Location of ditch discharge to final clarifier with respect to rotor (should be upstream of rotor and far enough upstream of influent and return sludge to prevent short-circuiting). The discharge weir shall be designed to withdraw mixed liquor from below the channel water surface such as through the use of a baffle in front of the weir. A baffle in front of the ditch overflow weir will prevent decanting clear liquid from the ditch. A floating draw off weir should be encouraged. ______________________

Water level control in aeration channel (should be adjustable weir). ______________________

Weir length of channel discharge (use maximum raw flow plus maximum recirculated flow to determine weir length to prevent excessive rotor immersions. ______ FT ______________

Type of access to rotor for maintenance (should be at least walkways located upstream of rotor and located to prevent spray from rotor on walkway). ______________________

Horizontal baffles should be located within 15 feet downstream of brush aeration in channels with water depth over six feet. This provides proper mixing of the entire depth of the basin. Channels using jet aeration may be deeper and should have a water depth as recommended by the manufacturer. ______________________

In a single channel, rotor drive assembly should be on outboard side. ______________________

Drive and gear assembly elevated out of water for safe and easy access. ______________________

Equipment requiring normal maintenance should be housed and /or heated. ______________________

PROTECTIVE AND SERVICE FACILITIES

Ladders, hoists and walkways provided for easy access to equipment, pipes and valves for normal operation, routine maintenance or repair.
Tanks with freeboard greater than 18" shall provide rails, steps or other means of egress from each aeration tank. Additional freeboard or windbreak may be necessary to protect against freezing or windblown spray. Walkways should be located and /or designed to preclude spray and foam from freezing on their surfaces. Locating walkways away from or sufficiently elevated, above water levels and sing grated walkways are examples of methods used to prevent ice build up. ____________________

Hoisting equipment available for blowers, pumps and other heavy equipment ____________________

Adequate remote and local controls provided for mechanical equipment, especially waste sludge pump, return sludge control, sludge loadout, etc. ____________________

Chlorination flexibility provided for chlorination of return sludge. ____________________

Laboratory equipment to be provided, should be MLSS, settlemeter, settleabilty and dissolved oxygen. Consider minimum "Process Control" Tests. ____________________

Railings or other protective walls shall be placed around all aeration tanks. ____________________

List safety equipment to be provided near aeration tanks and clarifiers. Should include safety vests, lifelines, and rings and safety poles. ____________________

Sufficient lighting should be provided to permit safe working conditions near aeration tanks and clarifiers at night. ____________________

Minnesota Pollution Control Agency

Water Quality

Wastewater Technical Review and Guidance

DISINFECTION SYSTEM – REVIEW CHECKLIST

Water/Wastewater/#5.22, May 2001

FACILITY NAME ______________________ DATE ______________________

CONSULTING ENGINEER ______________________ SITE INSPECTION (DATE & INSPECTOR) ______________________

PLANNING OR DESIGN PHASE ______________________

Type of Disinfection
(fill out appropriate section below.) ______________________

Pilot study
(for disinfection, other than chlorine, pilot/full scale studies are recommended) ______________________

A. <u>Chlorination</u>

1. Contact Basin Design/Detention Time ______________________
   (>15 minutes at peak hour flow)

   a. L/W ratio *(minimize short-circuiting)* ______________________

   b. volume ______________________

   c. peak hourly flow rate ______________________

   d. detention time at peak flow rate ______________________

2. Dosage ______________________
   (Normal domestic sewage effluent guide)

Treatment Type	*Dosage(mg/l)*
A. Trickling Filter	*10*
B. Activated Sludge	*8*
C. Tertiary Filtration	*6*
D. Nitrified	*6*

Contents:

wq-wwtp5-22

Minnesota Pollution Control Agency, 520 Lafayette Road North, St. Paul, Minnesota 55155-4194
(651) 296-6300, toll-free (800) 657-3864, TTY (651) 282-5332 or (800) 657-3864

3. Chlorination Equipment and Supply ______

   a. point of application ______
   (preferably before contact chamber, such as influent pipe prior to chamber to provide mixing)

   b. stand by equipment/spare parts ______
   (capacity to replace largest unit during shutdown with spare equipment to replace parts subject to wear or breakage)

   c. are scale provided? ______

   d. mixing ______
   (positively mixed effected in 3 seconds with flash mixer or turbulent flow)

   e. liquid chlorine storage ______
   (non-metallic storage with secondary containment, storage limit approximately one month)

4. Provisions for sludge removal ______
   (duplicate tanks, mechanical scrappers or vacuum cleaning equipment)

5. Location and space ______
   (room near point of application, near ground level, sufficient storage area)

6. Safety in chlorine supply room ______

   a. isolated from other work areas ______
   (gas tight, separate building or entry only from outside if connected with other building; include panic hardware;

   b. fire protection ______

   c. floor drains and ventilation separated from other plant/buildings ______

d. Inspection window indoor and separate wall. ______
(door window mandatory, separate wall window preferred)

e. temperature control ______
(minimum temp. for disinfection equipment = 60°F; protection against excess heat; cylinder storage near room temp.)

f. ventilation ______
(forced mechanical ventilation which provide one complete air exchange/min. when the room is occupied)

1) scrubbing provided or required by local Fire Marshall? ______

2) fans and lights operated manually from outside ______

g. electrical ______

1) fans and lights operated manually from outside ______

2) sealed electrical outlets provided ______

h. chlorine leaks ______
(automatic detection and alarm system)

i. cylinders fastened to wall or otherwise stabilized ______

7. Sampling point provided? ______

B. <u>Dechlorination</u>

1. Is dechlorination required? ______

2. Dosage (sulfite (SO_2)) __________

Dechlor. Chem.	*Theorectical Required per (mg/l) Chlorine*
Sulfur dioxide (gas)	*0.9*
Sodium meta bisulfite	*1.34*
Sodium bisulfite	*1.46*

Note: excess sulfur dioxide may consume oxygen. Consider reaeration.

3. Equipment __________
   Multiple units should be considered to compensate for high and low flow ranges. This will minimize excessive application. Feed rates should be based on chlorine residual rates

4. Detention time __________
   Minimum 30 seconds at peak hourly flow or max. rate of pumpage

5. Standby equipment/spare parts __________
   Similar to chlorination above

6. Storage room/safety __________
   Similar to chlorination above

7. Sampling point provided __________

8. Is mixing provided. __________

C. Ultraviolet

1. Quality of effluent __________
   (at least 65% ultraviolet radiation transmittance at 254 nanometers and BOD and suspended solids always less than 30 mg/l)

2. Lamp cleaning provided for? __________

D. Ozone

1. Evaluate on a casy-by-case basis. Design standards, operating data and experience with this process are not well established. __________

Minnesota Pollution Control Agency

Guidelines for wastewater-treatment facilities during a flood

Fact sheet #wq-wwtp7-01, April 2001

Flooding poses many problems and challenges for municipalities and industries. Among these are the protection and continued operation of wastewater-treatment facilities (WWTFs). This fact sheet is intended as guidance for WWTF operators in preparing for potential floods and working to recover operation after their facilities have been affected by a flood.

Avoid bypassing if possible

Obviously, every WWTF operator wants to avoid bypassing untreated wastewater at all times, if possible. However, sometimes it becomes necessary for a number of reasons, including:

Power outages — Cities and industries can avoid or minimize bypassing due to power outages if generators are available to provide emergency electricity to power lift pumps. For those facilities that have emergency generators available, it is important to start the generators ahead of the need to make sure they will operate properly in emergency situations. It is also prudent to plan ahead for fuel for the generators.

Lift pump failures — The most common problem during floods is failure of overworked lift pumps. If possible, lift pumps should be inspected and serviced ahead of a potential flood so they will operate at peak efficiency as flows begin to rise. If your lift pumps have had recurring problems with specific parts or components in the past, it might be wise to have spare replacement parts on hand to repair pumps in emergency situations.

Excess flows — As floodwaters rise, sump pumps begin running full time and water may find its way into the sanitary collection system through leaky manholes, open connections or cracked and leaky collection pipes. Lift pumps can be overwhelmed and unable to transport all of the flow to the treatment system. Sump pumps should be disconnected from the sanitary system. If possible, some manhole cover sealing can be done. Open connections should be located and sealed.

Treatment system problems — Pond systems are normally very full in the spring. Extra water from flooding can raise water in ponds to critical levels. In this situation, it is preferable to open the discharge structure to allow water to leave the pond system at the same rate it is entering. Be cautious of allowing water to continue to accumulate in the ponds to the top of the dikes. The clay dike core or vinyl liner does not extend all the way to the dike top. Maintaining water above the core or liner can lead to catastrophic dike failure.

wq-wwtp7-01

Minnesota Pollution Control Agency, 520 Lafayette Rd. N., Saint Paul, MN 55155-4194
(651) 296-6300, toll-free (800) 657-3864, TTY (651) 282-5332 or (800) 657-3864

If you must bypass

If bypassing is the only alternative, remember that your first obligation is to contact the Minnesota State Duty Officer at (800) 422-0798) or (651) 649-5451. If you need assistance or advice from the Minnesota Pollution Control Agency (MPCA), you can inform the Duty Officer or contact the MPCA directly.

Your second obligation is to discontinue the bypass as soon as possible. If a bypass is expected to last for more than a day or two, contact your MPCA representative and keep him or her informed of your problems and progress.

Other potential problems

Fuels in sewers —A number of mishaps can occur during floods that can lead to spills of fuel, which enter the WWTF collection system. Significant amounts of fuels can cause many problems for treatment systems whether they are pond or mechanical systems. If a fuel spill enters the collection system, call the Duty Officer immediately. Determine the origin of the spill and approximate volume. Fuels are lighter than water and they will float, so it is sometimes possible to contain the spill in the wet well by adjusting float levels or running the pumps manually, so the water lever does not pump all the way down. If the spill does reach the treatment system, the type of system and amount of fuel will determine how serious the problem becomes. Fuels can completely kill beneficial bacteria in a mechanical system and it may need to be "reseeded" after the fuel has passed. Pond systems can handle some fuel without major problems. The fuel should be contained in a single pond, where it may be recoverable with proper equipment.

Flooding around a pond system —During the 1997 flood, many pond systems in the Red River Valley had floodwaters rise to within a foot or two of the tops of the dikes. One of the problems this poses is that dikes become saturated from inside and out. This can weaken dikes and, as floodwaters recede, may lead to dike failure from the pressure of the water in the ponds pushing against saturated dikes. During the 1997 flood, pond operators were advised to open gates and valves and allow water in the ponds to go down with the receding floodwaters. This should only be done when absolutely necessary and after consultation with MPCA staff.

Plan ahead.

Finally, if you believe you may have to face flooding problems, contact MPCA staff to discuss potential problems. Make sure you have any emergency numbers you may need in a handy place and make sure that other staff who may assist you are aware of these numbers and of the need to contact the Minnesota State Duty Officer or others for assistance.

For further assistance

If you need further assistance, call the Water Quality Helpline at the MPCA in St. Paul at (651) 296-7162 (or call the MPCA toll free at 800/657-3864 and ask for the Water Quality Helpline).

Assistance is also available from the following MPCA offices outside the Twin Cities metro area:

Brainerd (218) 828-2492
Detroit Lakes (218) 847-1519
Duluth (218) 723-4660
Mankato (507) 389-5977
Marshall (507) 537-7146
Rochester (507) 285-7343
Willmar (320) 214-3791

To report a bypass or a fuel spill of more than five gallons, call the Minnesota State Duty Officer at (800) 422-0798 or (651) 649-5451.

N.1.1.6 *Minnesota's Wastewater Treatment Facility Operation and Maintenance (OM) Manual Guidelines*

Minnesota Pollution Control Agency

Water Quality

Wastewater Technical Review and Guidance

Wastewater Treatment Facility Operation and Maintenance (OM) Manual Guidelines

Water/Wastewater/#7.00, February 2000

Facility Name ____________________ Date ____________________

Consulting Engineer ____________________

I. Introduction

The operation and maintenance (OM) manual is a very valuable tool in the operation and maintenance of every wastewater treatment facility. The OM manual serves two main functions; first, as a text to be used during start-up training by the consultant and second, to provide assistance to the operating staff in continuing standard operating and maintenance procedures after start-up.

Past studies have shown OM manuals often to be inadequate for operating personnel, a common problem being practicality and overall approach. If these manuals are to benefit treatment facility operation, they **must be oriented toward the treatment plant operators.** The OM manual must be written for the entire operating staff and contain information pertinent to the day-to-day operation of the treatment facilities.

II. Definitions/Requirements

A) OM Manual Definitions - The OM manual is referred to as either the "final" OM manual or the "revised final" OM manual.

The **final OM manual** is given to the project owner prior to the actual start-up of their facility. The **revised final** OM manual is the final OM manual after it has been revised as a result of the first year of operation.

wq-wwtp7-00 **Minnesota Pollution Control Agency, 520 Lafayette Road North, St. Paul, Minnesota 55155-4194**
(651) 296-6300, toll-free (800) 657-3864, TTY (651) 282-5332 or (800) 657-3864
This material can be made available in alternative formats for people with disabilities.
Printed on recycled paper containing at least 20 percent fibers from paper recycled by consumers.

B) Final OM manual -The final manual will be reviewed for content and completeness and if found satisfactory will be approved as a final O & M manual. The final manual must be complete enough to be used as a textbook for the operators during start-up training. In other words, the final manual should be void of spelling and typographical errors and must include all diagrams, figures and tables, page numbers, all design data and parameters, etc. It must be essentially a complete manual except for very specific information that may not be available. The final manual must be submitted for approval at least 45 days prior to initiation of operation. **The actual date for initiation of operation will not be approved until the final manual has been approved.**

C) Revised Final OM Manual - The revised final O & M manual must be submitted for review at least 30 days prior to the completion of the performance certification period (eleven months after initiation of operation). The revised final manual is an updated final manual that includes revisions based on actual plant operation. The revised final manual must be prepared by a cooperative effort between the Consultant and the treatment facility operating staff. To assure proper input by the operating staff, the revised final manual must be submitted with a cover letter that includes the signatures of the treatment plant superintendent and/or principal operator and the person(s) responsible for preparing the manual.

The project certification report will not be approved unless the revised final OM manual has been approved.

III. OM Manual Content

These guidelines present a suggested format for OM manual preparation. Consultants are encouraged to improve upon these guidelines to accommodate unique treatment plant requirements. It is not the intent of these guidelines to restrict or stifle initiative. Format other than these guidelines will be accepted. However, the chapters as presented in these guidelines must still be provided. The OM manual should not be written in this format just to satisfy the Minnesota Pollution Control Agency (MPCA) review. The overlying reason for writing the OM manual must be to provide OM instructions to the plant operator(s).

Operator input is required in development of the revised final OM manual. Operator input is also suggested in preparing the final manual as this will make the manual more amenable to the operator and easier to revise. For example, the operators may have certain procedures, records, lab forms etc., that they prefer which should then be incorporated in the final manual. The operations staff may prepare many parts of the OM manual. For example, when the operators implement their maintenance programs as per the OM manual instructions, they would be in a good position to also develop the table summarizing the maintenance schedule of all equipment for the revised final manual.

It is suggested that the OM manual be bound in 3-ring binders to facilitate easier revision at a later date.

A) STABILIZATION POND SYSTEMS (with or without land application)

All stabilization pond operation and maintenance manuals (with or without spray irrigation) must include a copy of the latest edition of the MPCA Stabilization Pond Manual along with the following additional information:

1. Description of pond system, including a flow diagram, design summary, and control structure diagrams.
2. Copy of the NPDES permit and plan/specification approval letter, including any special permit requirements.
3. Discussion on actual operation of the pond system.
4. Example of how to calculate discharge volumes (for various flow rates) specific to the pond system.
5. Discussion of NPDES permit requirements for "acceptable and problem discharge periods".
6. Recommended qualifications, staffing recommendations, and operator certification.
7. Minnesota OSHA confined space entry requirements, along with how this applies to this system.

For the most efficient preparation of an OM manual for a pond system, we recommend that the MPCA pond manual be used verbatim as a base with specific information concerning the above items to be added on different colored paper to each chapter or consolidated at the end of manual. The OM manual should begin with an explanation page outlining manual organization used and explaining the difference between the MPCA pond manual and the specific sections of the OM manual.

B) MECHANICAL PLANTS

The following is only a recommended format. The consultant may improve upon it or may use any other format. However, all OM manuals must incorporate all of the chapters outlined in the following recommended guidelines.

The final manual should be complete enough to be used as a textbook for start-up training and plant start-up, including suggested operational criteria to be used as a guideline in attaining optimum treatment efficiency as soon as possible. The final manual should include design operating parameters such as unit efficiencies, operating criteria, suggested design or process control test ranges, etc. The operational areas should be a series of simplified instructions for plant personnel to follow. The revised final manual should include recommended operational values and modes of operation based on actual plant operation and data obtained during the one year performance certification period.

Chapters I, II, VII, IX and X of the final manual should be written in final form as little change is expected in these chapters between the final and revised final OM manuals. However, this does not preclude any necessary changes from being made. **The OM manual must be written to instruct the operations staff how to operate the treatment facilities. It should not be written just to follow this recommended format.**

The format recommendation is as follows:

CHAPTER I: INTRODUCTION

- Introduction
- Operation and maintenance responsibility described for the operators and for the city or governing body.
- General description of plant type and all units including flow diagram and plant layout with major piping included.
- Design summary of the overall treatment facility, including plant and unit influent design data, number of units, size (s), design and design peak loading, design performance data, etc. This information may be provided in Chapter III where each individual unit is discussed.
- Discussion of treatment requirements/NPDES effluent limitations.

CHAPTER II: PERMITS

- Reporting procedures for spills of raw or inadequately treated wastewater and toxic/hazardous wastes including NPDES permit requirements to report and explain effluent violations.
- Copy of the treatment facility plan and specification approval letter from the MPCA and discussion of any special requirements.
- Copy of the NPDES Permit, including discussion of any special requirements in the permit such as bypass authorization, special provisions, compliance schedule, pretreatment requirements, etc.

CHAPTER III: OPERATION AND CONTROL OF WASTEWATER TREATMENT

- Individually discuss each category of units and auxiliary systems associated with each unit such as bar screens, grit chambers, primary clarifiers, sludge pumps, auxiliary power, alarms etc. including the following information (if applicable):

NOTE: Many of these sections need not be very detailed, if reference can be made to other areas of the OM manual. For example, the control/lab tests and maintenance section may reference other chapters of the OM manual.

a. Purpose of the equipment or unit(s), what does the unit do?

b. Equipment or unit description, including schematics, how does it accomplish its purpose? The description should include sizes and dimensions and trace all flows (wastewater, sludge, recirculation, etc.) through the unit and comment on how the wastewater or sludge is altered by the unit. The theory behind how the process works should also be included if the process is not included in the MPCA Wastewater Treatment Technology manual. It is suggested that the MPCA Wastewater Treatment Technology manual be incorporated into the final OM manual.

c. Relationship to adjacent units - discuss how the operation of each treatment process will affect or be affected by adjacent units.

d. Operation - including start-up, normal operation, alternate modes of operation, seasonal operation, bypass operation, shutdown and draining procedures. Normal operation discussion should include initial operational parameters and a systematic approach to attaining these parameters with a detailed step-by-step procedure for making adjustments and receiving feedback through the use of control tests and physical measurements. Operator application of concepts and testing to process control must be emphasized. This section should also include valve positions for each mode of operation, when to use the unit or alternative mode of operation, flow rates, sludge concentrations, expected results or efficiencies, operation during industrial shock loads, I/I, etc.

e. Controls - such as flow controls (flow to the unit, recirculation, air, or sludge), electrical controls and manual controls, alarms, etc.

f. Unit or related equipment diagram or schematic(s) including valve location.

g. Potential operational problems and solutions, such as a troubleshooting guide.

h. Maintenance - specific maintenance may be discussed here in the revised final manual. Reference may also be made to manufacturer's manuals, if the manufacturer's manual is specific for that piece of equipment and can be referenced to the appendix.

CHAPTER IV: OPERATION AND CONTROL OF SLUDGE HANDLING

- Individually discuss each category of units involved with handling sludge such as sludge holding, thickening or treatment units, sludge pumps, sludge disposal, etc. including the following information if applicable:

a. Purpose of the unit

b. Equipment description - the description should include schematics in addition to sizes and dimensions of the units and trace all flows through the unit and comment on how the character of the sludge is altered. Discussion of theory for aerobic or anaerobic digestion and land application of sludge will be provided through a chapter from the MPCA Wastewater Treatment Technology manual may be incorporated into the final OM manual. All other sludge treatment or handling processes must include a discussion of theory.

c. Relationship to adjacent units - discuss how the operation of each treatment process being considered can affect other units.

d. Operation including start-up, normal operation (design condition), alternate modes of operation, bypass operation, draining procedures. Each discussion should include valve positions, flow rates, sludge concentrations, expected results or efficiencies, use of control tests, etc.

e. Controls such as flow controls (flow to unit, sludge flows, air rates, etc.), electrical controls and manual controls

f. Unit diagram including valve locations

g. Potential operational problems and solutions

h. Maintenance

In addition, the sludge-handling chapter should contain a copy of the approved biosolids management plan or adequate information to instruct the operating staff to be able to develop a biosolids management plan. **A copy of the MPCA Land Application of Biosolids Manual must be included.**

CHAPTER V: PERSONNEL

- Outline manpower requirements and recommended qualifications of each such as training, experience, skills and certification eligibility requirements. The revised final manual should include any recommended staffing changes, training needs, and etc. based on the first year of actual operation.

- Discuss pertinent aspects of operator certification as they apply to this facility, including wastewater and waste disposal (land application of biosolids) certification requirements.

CHAPTER VI: PROCESS CONTROL AND LABORATORY TESTING

- In the past, OM manuals have been quite inadequate in the area of process control. One suggested method of presenting this information is to review each unit individually and discuss the laboratory tests and test results as they apply to that unit.
- Discuss testing requirements (control testing, NPDES requirements, etc.) and the purpose of testing; this should include control testing and taking physical measurements. Also discuss how to record tests for easy use such as trend charts.
- Discuss proper sampling procedure.
- Give instruction for conducting tests or taking physical measurements, **including interpretation of test results or measurements (i.e. expected range, what high or low values may indicate, etc.) for each application of the test and how the operator should apply the test results to process control**. When discussing test procedures, do not reprint "Standard Methods", but reference when applicable. Give instructions in the OM manual for only those tests not found in "Standard Methods" or other approved manuals.
- Outline quality control program for NPDES and control tests conducted on-site.
- Sample laboratory worksheets

CHAPTER VII: RECORDS

- Discuss and provide examples of records for daily operations, equipment, costs, personnel, lab, etc. that will be of use to the operations staff.
- **Discuss and provide example of MPCA monthly operation report requirements.**
- Discuss and provide example of an annual wastewater report and a budget.
- Discuss and provide example of biosolids landspreading annual report form.

CHAPTER VIII: MAINTENANCE

NOTE: The general discussion should be complete in the final manual including the instructions on developing a maintenance program. As a result of this chapter, the operating staff should be able to develop, implement and utilize a maintenance program including a system to schedule, conduct and record all maintenance. During the first year of operation, the maintenance program outlined in the final O M manual is implemented by the operations staff, with assistance from the consultant. The revised final manual then should include a table developed during start-up summarizing the maintenance schedule of all equipment on a weekly, monthly, etc. basis including the type of lubricant.

- **Outline purpose of maintenance program**

- Discuss basic features of maintenance program and management system to include:

    a. Maintenance procedures

    b. Equipment maintenance record system with instructions in starting and carrying out such a record system

    c. Planning and scheduling, how to utilize the record system

    d. Inventory system

    e. Special tools and their use

    f. Warranty provisions including specific dates when the warranties run out (to be in the revised final O M manual or incorporated into the maintenance record system).

    g. Maintenance cost and budgeting system

    h. Summary of maintenance schedule of all equipment on a weekly, monthly, etc. basis including the type of lubricant, etc. (to be in the revised final O M manual).

CHAPTER IX: EMERGENCY OPERATING AND RESPONSE PROGRAM

- Vulnerability analysis, an analysis of the treatment facility with respect to the emergency situations that would be most vulnerable to its operation, i.e. flooding, tornado, fire, personnel strike, power outage, etc.

- Recommend equipment list to handle emergency situations.

- Emergency operating plan for maintaining operation during the listed emergencies including mutual aid recommendations. Included also must be NPDES permit instructions for notification of the State Duty Officer.

CHAPTER X: SAFETY

- General safety discussion including hazards of sewers, mechanical equipment, explosion, bacterial infections, chlorine, oxygen deficiencies, lab, electrical etc.
- General discussion and use of safety equipment.
- **Safety relating to the specific areas and necessary safety equipment.**
- **Discussion of OSHA Confined Space Requirements.**

CHAPTER XI: APPENDICES

- The appendices should be a bound manual with Table of Contents, tabs or other indexing system, page numbers, etc., for easy reference. The appendices should include at least all the following applicable information:

  a. schematics

  b. valve indices

  c. sample forms

  d. NPDES Permit/plan and specification approval letter

  e. List of manufacturers manuals: the manuals need not be included in the OM manual, but a list must be provided referencing their location.

  f. Copy of the MPCA Discharge Monitoring Report forms and instructions

  g. Copy of wastewater certification rules, biosolids rules, landspreading site approval form, biosolids landspreading annual report form.

  h. List content of Reference Library

IV. SUBMITTING AN O&M MANUAL

One copy should be submitted for review purposes. Upon approval, two copies of the OM manual will be needed. One copy will be for the MPCA files and the other will be included with the approval letter.. The final manual should be submitted at least 45 days prior to the initiation of operation date The revised OM manual must be submitted for approval at least 30 days prior to the end of the one-year certification period (11 months after the initiation of operation date). If no revisions are needed, a letter must be submitted stated this.

As a reminder, the actual date for initiation of operation will not be approved until the final manual has been approved. Also, the project certification report will not be approved unless the revised final OM manual has been approved.

Should you have any questions, feel free to contact Gene Erickson at (800) 657-3864.

N.1.2 Virginia

The Virginia Sewage Collection and Treatment (SCAT) Regulations require a technical evaluation and approval of proposals, reports, plans, and specifications submitted in support of the issuance of permits to construct and operate sewage collection and treatment facilities. These regulations are implemented by the Virginia Department of Health (VDH), Division of Wastewater Engineering (DWE). The DWE program ensures that technical assistance is provided to owners and their consultants to develop the most cost effective technology that can be reasonably operated to provide reliable performance in compliance with permit requirements. The DWE staff negotiates an acceptable design with the consultant and owner in accordance with the SCAT Regulations. During this process, small communities will receive the benefits of the statewide technical expertise provided by the VDH staff. In addition, equipment manufacturers will be provided with a uniformly regulated process on which to base competitive bids for construction contracts.

Currently an average of over 1,000 sewage collection, and treatment projects, representing nearly 250 million dollars in construction costs, are submitted to DWE annually, for a technical evaluation. The overwhelming majority of such submissions originate from local governments through their private consulting engineer who presents the proposed design at a preliminary engineering conference with the DWE staff. Some errors or omissions are usually identified on all submitted technical documents. Correction of misleading information used for contractor bids results in eliminating potential problems prior to initiating construction of the sewerage system or treatment works, and thus saves substantial amount of dollars, otherwise expended by the local governments that own and operate several thousand sewage collection and treatment systems, transporting and processing hundreds of million gallons of raw sewage each day.

The SCAT Regulations establish new permit requirements in section 12 VAC 5-581-110 and provide for waivers of formal permits for small sewage collection and treatment systems. The proposed permit issuance procedures(12 VAC 5-581-130) are much more flexible than the procedures specified in the 1977 Sewerage Regulations. A General Permit will be available through the proposed regulations (12 VAC 5-581-280) for local approval of projects involving the expansion of sewage collection systems.

A major difference between the SCAT Regulations and the previous Sewerage Regulations involves construction project evaluation procedures and construction permits. The SCAT Regulations will allow issuance of a construction permit without a formal technical evaluation of design documents(plans and specifications), based upon the information provided at the preliminary engineering conference and the submittal of a signed statement by the responsible design consultant that the design will meet the minimum required standards.

Reliability provisions are based on a measurement of the ability of a component or system to perform its designated function without failure or interruption of service. Overflow criteria, such as a period of discharge, are utilized solely for the establishment of reliability classification for design purposes and are not to be construed as authorization for, or defense of, an unpermitted discharge to state waters. The reliability classification will be a major consideration for discussion at the preliminary engineering conference described in this chapter. Pump stations associated with, but physically removed from, the actual treatment works may have a different classification than the treatment works itself. The reliability classification shall be based on the water quality and public health consequences of a component or system failure. Guidelines for classifying sewerage systems and treatment works are as follows:

Reliability Class I. Sewerage systems or treatment works whose location, or discharge, or potential discharge (i) is sufficiently close to residences, public water supply, shellfish, or recreation waters; (ii) has a volume or character; or (iii) for which minimal dilution of 10 to 1, receiving water volume to discharge volume, based on permit flow values is not provided year round, such that permanent or unacceptable damage could occur to the receiving waters or public health if normal operations were interrupted.

Reliability Class II. Sewerage systems or treatment works whose location or discharge, or potential discharge, due to its volume or character, would not permanently or unacceptably damage or affect the receiving waters or public health during periods of short-term operations interruptions, but could be damaging if continued interruption of normal operation were to exceed 24 hours.

Reliability Class III. Sewerage systems or treatment works not otherwise classified as Reliability Class I or Class II. Unless designated as applying to a particular reliability class, all requirements specified in the regulations apply equally to all reliability classes.

The objective of achieving reliability protection is to prevent the discharge of raw or partially treated sewage to any waters and to protect public health by preventing backup of sewage and subsequent discharge to basements, streets and other public and private property.

Provisions for continuous operability of pumping stations shall be evaluated in accordance with the appropriate reliability classification.

Independently operated essential equipment or components of sewerage systems and treatment works shall be provided with sufficient duplication or alternative operation so that the average daily design flow may be transported, stored, treated or otherwise managed in accordance with reliability requirements with the largest component out of service. Sufficient spare parts to ensure continuous operability of essential unit operations and equipment shall be kept in a central storeroom located at the treatment works or at other readily accessible locations, and the minimum quantities shall be in accordance with the operation and maintenance manual. The need for spare parts should be determined from review of manufacturer's recommendations, evaluation of past maintenance requirements, etc. A spare parts inventory shall be included in the operation and maintenance manual. The inventory shall list the minimum and maximum quantities of the spare parts to be kept on hand, the equipment in which they are used, their storage location, replacement procedures and other pertinent information. Plans and specifications for a pump station submitted to the Division proposing to use portable equipment to meet continuous operability requirements shall be accompanied by a completed "Portable Equipment for Sewage Pump Stations" form.

For Class I reliability, the alarm system shall monitor the power supplies to the station, auxiliary power source, failure of pumps to discharge liquid, and high liquid levels in the wet well and in the dry well, and shall include a test function. An on-site audio-visual alarm system shall be provided such that each announced alarm condition is uniquely identified. In addition, provisions shall be made for transmitting a single audible alarm signal to a central location where personnel competent to receive the alarm and initiate corrective action are either: (i) available 24 hours per day, or (ii) available during the periods that flow is received at the pump station. A sign indicating notification procedures (responsible persons, telephone numbers, etc.) to be followed in case of alarm actuation shall be displayed conspicuously.

Classes II and III. For Class II or III reliability, the alarm system shall monitor high liquid levels in the wet well. An on-site audio-visual alarm signal shall be provided. A sign indicating notification procedures (responsible persons, telephone numbers, etc.) to be followed in case of alarm actuation shall be displayed conspicuously. A backup power supply, such as a battery pack with an automatic switchover feature, shall be provided for the alarm system, such that a failure of the primary power source would not disable the alarm system. A backup power supply for the alarm system should be provided for a Reliability Class I facility with dual electrical feed sources. Test circuits shall be provided to enable the alarm system to be tested and verified to be working properly.

The treatment works design shall provide for satisfactory operation during power failures, flooding, peak loads, equipment failure, and maintenance shut-down (in accordance with the requirements of the appropriate reliability class). Such design features include: (i) additional electrical power sources; (ii) additional flow storage capacity; and (iii) additional treatment unit operations, that provide for alternate operation in accordance with the issued certificate permit requirements.

For Class I Reliability, two separate and independent sources of power feed shall be provided. Each source shall be capable of maintaining continuous treatment works operation at peak design flow during power failures, flooding, or equipment malfunction. Certain Reliability Class I treatment works for which it is feasible to shut down or discontinue treatment works operation during periods of power failure without bypassing or violating effluent limitations may be exempt from the alternate feed requirement.

In combination with provisions for electrical power reliability, the use of flow storage and additional unit operations should be evaluated. Additional flow storage capacity should provide up to a 24-hour detention of the peak design flow. Additional unit operations could involve chemical clarification, filtration, additional disinfection capacity, or use of natural treatment technology for enhancing effluent quality.

An audio-visual alarm system to monitor the condition of equipment whose failure could result in a bypass or a violation of effluent limitations shall be provided for all treatment works. Alarms shall also be provided to monitor conditions which could result in damage to vital components.

APPENDIX O

CASE STUDY INFORMATION

O.1 Case Study Reponses

Responses to the project team's case study questionnaire are provided in Table O-1.

Table O-1. Case Study Results.

Case Study Questions/ Responses Received	City of Santa Barbara El Estero Wastewater Treatment Plant Santa Barbara, California	City of Toronto, Ashbridges Bay Treatment Plant, Toronto, Ontario, Canada	Orange County Utilities, South Water Reclamation Facility, Orlando, Florida,	Western Lake Superior Sanitary District, Duluth, Minnesota
Which design redundancy practices are most effective?	1. Chlorination, Dechlorination monitoring equipment. 2. Auxiliary Power Generator 3. Influent Pump	Redundancy practices that are applied to the critical process components are most effective. Critical process components must be identified and sufficient redundancy provided to ensure continuous service. This is site specific.	The design redundancy practices that eliminate or drastically reduce permit violations. Alternative operational control.	Sufficient redundancy to allow low staffing levels and insure backups for critical components and systems.
What is the premise for these practices?	1. Inspected SB 709 – mandatory penalties 2. Improves reliability of the treatment power	The premise is to deliver a reliable yet cost competitive service based on a sound preventative maintenance program. Sound preventative maintenance programs reduce the need for redundancy.	To expand a window of opportunity to meet Regulatory requirements, to cut operational cost by reducing down time.	A proper level of automation is economically supported by an appropriate level of backup.
Why do these design practices work?	Reliability enforcement	The design practices work largely due to implementing proven technology.	Operations and Maintenance staff usually do not realize the impact of noncompliance parameters under Regulatory agencies. Redundancy design practices ensure that some source of operational alternative is applied.	Most stakeholders have needs met. • Operations – operations flexibility • Management – reliability and efficiency • Owners – economy
What level of design redundancy is required for optimal performance?	99% redundancy required for every piece of equipment needed to operation of the plant.	Design redundancy should be provided to suit the critical process components only. All non-critical	Optimal performance is usually associated with automatic computer controlled systems that	A level that delivers the values the stakeholders say are important. It is also dependent upon technology

Table O-1. Case Study Results (continued).

Case Study Questions/ Responses Received	City of Santa Barbara El Estero Wastewater Treatment Plant Santa Barbara, California	City of Toronto, Ashbridges Bay Treatment Plant, Toronto, Ontario, Canada	Orange County Utilities, South Water Reclamation Facility, Orlando, Florida,	Western Lake Superior Sanitary District, Duluth, Minnesota
		Components can have minimal redundancy.	Function with little or no human intervention with backup features.	in place, asset condition, and competence of staff.
What are the cost benefits of applying redundancy practices?	Avoid or prevent high cost mandatory fines such as $3,000 per discharge violation.	Redundancy practices can only be cost effective if the technology applied is proven to be reliable. Excessive redundancy lends itself to excessive capital costs while quite possible increasing irrelevant maintenance costs associated with frivolous redundancy.	As with most systems, cost is a benefit when the system works. Preventing monetary fines and environmental impacts saves money. There are no cost benefits from failure of a redundant system.	• Improved flexibility • Scheduling advantages • increased maintenance planning • reduces ad hoc operations/activity • minimizes startup/shutdown of process • saves time/process chemicals
How can these practices be implemented?	Budgeted and good planning.	Redundancy practices can only be implemented effectively if the treatment process is understood. The critical process elements must be identified for sound implementation.	Design build teams consisting of front line workers and the design staff have paved new roads for redundancy concepts. Again, cost directly impacts an Owner's ability and willingness to reduce or eliminate worse case scenario failure cost.	Most efficient in original design, can be added as part of plan to address sub-optimal performance, a utility with a performance culture could add to continuous improvement plans.
Can you provide the documentation or a reference to the documentation of your in-house requirements/ standards for redundancy design practices?	N/A. There are no existing requirements.	The City does not have a document for in-house redundancy design practices. However, the Ministry of Environment and Energy (MOEE) has a dated design guideline for wastewater plant design in Ontario and the where relevant WEF	In August 1998, Orange County Utilities contracted Environmental Engineering firms to develop a Wastewater Treatment Plant Automation Plan. The scope of the plan was to evaluate current plants within the County and make	Other than the Clean Water Act, these practices are very specific and need to be documented.

Table O-1. Case Study Results (continued).

Case Study Questions/ Responses Received	City of Santa Barbara El Estero Wastewater Treatment Plant Santa Barbara, California	City of Toronto, Ashbridges Bay Treatment Plant, Toronto, Ontario, Canada	Orange County Utilities, South Water Reclamation Facility, Orlando, Florida,	Western Lake Superior Sanitary District, Duluth, Minnesota
		"Design of Municipal Wastewater Treatment Plants" is utilized. Both of these documents can be obtained directly from the MOEE and the WEF.	recommendations for improvements.	
How do your in-house requirements/standards relate to the state or provincial requirements/ standards?	N/A	Being the largest facility in the Country, our standards tend to fall in line with standard design practices in the industry within North America.	In order to support the cost and needs for redundancy. In-house staff utilizes known practices that are proven for the scope of work to be accomplished and many times Orange County Utilities is the forerunner of new systems that have been evaluated by in-house Staff, Engineers and Consultants.	They were included in the plans and specifications provided with facility plan for recent biosolids project.
Is the state or provincial redundancy policy a regulation or a guideline?	Regulation	The province only provides guidelines and for the most part under the issuance of the Certificate of Approval, regulates standby power requirements and firm capacity requirements for pumping stations for prevention of overflows, spills and basement flooding. At Ashbridges, most components are designed to handle the average daily flow within the plant with	To meet State of Florida requirements to protect the Environment and Public Safety and Health. Engineering designs must be approved by the State through permit applications and operations Protocols.	I understand it as a guideline but whatever is proposed must be approved by the agency.

Table O-1. Case Study Results (continued).

Case Study Questions/ Responses Received	City of Santa Barbara El Estero Wastewater Treatment Plant Santa Barbara, California	City of Toronto, Ashbridges Bay Treatment Plant, Toronto, Ontario, Canada	Orange County Utilities, South Water Reclamation Facility, Orlando, Florida,	Western Lake Superior Sanitary District, Duluth, Minnesota
		the exception of the primary settling tanks. The inlet works and the primaries as well as the ability to chlorinate is designed to handle the peak flows through the plant.		
Was the redundancy policy followed during your last plant expansion?	No. Followed-up after SB 709 was implemented in the year 2000.	The redundant guidelines are always followed as part of facility expansion.	Current upgrades and new construction systems require redundancy in many of the features. When applicable and funds allow, systems that are not required are designed for redundancy if the cost can be recaptured in a short term (two to three years) or if it can reduce man hours either by actual time saved or reduced down time.	Yes
Can you provide a plant process flow schematic and a line drawing of your automation system?	Yes	This information is available and can be reviewed with plant staff. Line drawings of the automation system are only available for sub-processes at this point in time. The facility is in the midst of consolidating these under the Process Control Strategy currently being implemented. This implementation is scheduled to take place over the next 2 years.	These will be will be provided under separate mail.	Yes

Table O-1. Case Study Results (continued).

Case Study Questions/ Responses Received	City of Santa Barbara El Estero Wastewater Treatment Plant Santa Barbara, California	City of Toronto, Ashbridges Bay Treatment Plant, Toronto, Ontario, Canada	Orange County Utilities, South Water Reclamation Facility, Orlando, Florida,	Western Lake Superior Sanitary District, Duluth, Minnesota
Please identify the population served, the average design flow and the design peaking factor.	Population served: 94,000 Average Design Flow = 11.0 mgd Average Peak Flow = 16.0 mgd	Population served: 1.25 million, average design flow = 818,000 m^3/d, design peaking factor = approximately 3 times (applicable for inlet works including pumping stations, primaries and disinfection only)	Population served: approximately 160,000 (50,162 + customer connections). Average Design Flow: Facility permitted for 30.5 mgd Peaking Factor = 2.0	Population served: 140,000 46 mgd (annual avg. daily) Average Peak Flow: 86 mgd Peak Hourly Flow: 160 mgd
Can you identify process trains, equipment and instrumentation added as a result of the redundancy policy and what impact did each have on the capital cost?	Chlorine Residual Analyzers Emergency Auxiliary Power Generator – Fixed 2.0 Megawatts Portable Auxiliary Power Generator – 260 Kilowatts	Redundancy implementation through the design phase of expansions and upgrades tend to be integrated into the overall scope of the engineering assignment. As such, it would be difficult to extract such information. As far as the construction and equipment costs are concerned, the latter may de distinguished but will require a great deal of effort in reviewing the tender documents and will also require some degree of analysis. This question cannot be answered without some extensive analysis being undertaken on all contracts let over the past 50 years.	The upgrade of the SCADA system has a redundancy backup to safe guard operations control and to allow for backup data information. Major items within the structured treatment facility have redundancy, such as aerations blowers, pumps, controls and DCUs. Backup storage capacity ensures hydraulic flow either high and low is controlled. The major impact on cost appears to be directed to the time from which an evaluation and recommended designs are approved to the contract awards. The changing economy relates to millions of dollars in short periods of time.	A very rough estimate using a recent biosolids project of additional costs associated with redundancy would be 5-10% of the total project costs. This additional cost is for extra process units, piping and backup systems. Staff view this type of design as a standard. Plant staff challenge designers to justify any designs which they believe does not meet the redundancy values expected by standardization.

Water Environment Research Foundation

Table O-1. Case Study Results (continued).

Case Study Questions/ Responses Received	City of Santa Barbara El Estero Wastewater Treatment Plant Santa Barbara, California	City of Toronto, Ashbridges Bay Treatment Plant, Toronto, Ontario, Canada	Orange County Utilities, South Water Reclamation Facility, Orlando, Florida,	Western Lake Superior Sanitary District, Duluth, Minnesota
What impact has the implementation of the policy had on the following:				
System performance	Enhances reliability in treatment performance	Provides a reliable system with the capability of handling breakdowns with minimal disruption.	A more stable and reliable control of systems. Requires an increase in the staff's technical and educational abilities.	Third AMSA Gold Award
Regulatory compliance	Improve reliability in meeting NPDES discharge standards	Ensures that the facility complies with the Certificate of Approval requirements at all times.	All Orange County Utilities facilities have a proven record of Regulatory compliance. All of the facilities have received awards for meeting or exceeding performance standards.	No operational issues other than SSOs in collection system due to inflow and infiltration (I/I).
Operating and maintenance costs	Major Capital Improvements, construction pending that will improve performance and redundancy • Digester mixing • Thickening control pump station • Additional influent pump and motor installation	Optimizes the operating and maintenance costs.	Operating cost has stabilized in the last three budget years while the volume of produced reclaimed water has increased. CIP costs have gone up dramatically. This was expected due to the new technology and improvements Orange County Utilities needs.	Stable or decreasing user rates past 3 years.
Energy utilization	Electrical usage increased by 128%	Able to manage demand more effectively.	Energy costs have risen through economics not redundancy.	Received a power company grant for $141,000 for net reduction of approximately 700 kW demand.

Table O-1. Case Study Results (continued).

Case Study Questions/ Responses Received	City of Santa Barbara El Estero Wastewater Treatment Plant Santa Barbara, California	City of Toronto, Ashbridges Bay Treatment Plant, Toronto, Ontario, Canada	Orange County Utilities, South Water Reclamation Facility, Orlando, Florida,	Western Lake Superior Sanitary District, Duluth, Minnesota
Any other measurable parameters	Recent wastewater rate study revealed an initial increase in sewer rates of 10% the first year and 3% for the following 5 years.	None.	System designs and redundancy have allowed for reduced staff requirements.	Substantial reduction in emerging maintenance.
What one factor was of prime importance in your adoption of your current redundancy design practices?	Ability to produce cost efficient wastewater treatment for Santa Barbara residents and visitors, and comply with NPDES standards.	Understanding the treatment process as this is unique to each facility.	Efficiency and cost savings to meet regulatory requirements.	Need for reliability and predictability of operations and maintenance costs.
What do you consider to be the most important factor in your successful implemen-tation of redundancy design practices?	Meeting discharge bacterial levels. MPNs in order to properly protect the ocean and public health environment.	Proper consultation with operational and maintenance staff during the design of upgrades.	Lessons learned. We strive to prevent reoccurrence of failures that impact our operational goals.	Buy in staff – management – board. It is very consistent with all stakeholder values.
What redundancy design practices that are in use at your utility do you feel are unique or alternative to most common approaches?	Auxiliary Power Chlorination/ Dechlorination monitoring equipment.	N/A	Our ability to store reclaimed water for freeze conditions during the winter months to protect a billion dollar a year citrus industry. This feature directly impacts the local, state and national economy and possibly the world citrus market.	Most have come from other operations both public and private.

REFERENCES

Briggs, R. 1998. Instrumentation, Control and Automation of Water and Wastewater Treatment and Transport Systems. *Water Sci. Tech.* 37(12).

Corbitt, Robert A., et. al. 1989. *Standard Handbook of Environmental Engineers.* New York: McGraw Hill.

Daigger, G. T., and Buttz, J.A. 1992. *Upgrading Wastewater Treatment Plants*. Lancaster, PA: Technomic Publishing Co. Inc.

Health Education Services, a Division of Health Research, Inc. 1997. *Recommended Standards for Wastewater Facilities (10-State Standards)*. Albany, NY.

Institute of Electrical and Electronic Engineers, Inc. (IEEE). 1974–1981. *Standards that address Redundancy for Nuclear Power Plant Equipment and Systems*. New York, NY.

- 308-1980—Criteria for Class 1E Power Systems for Nuclear Power Generating Stations
- 379-1977—Application of the Single Failure Criterion to Nuclear Power Generating Station Class 1E Systems
- 384-1981—Criteria for Independence of Class 1E Equipment and Criteria
- 387-1977—Criteria for Diesel Generator Units Applied as Standby Power Supplies for Nuclear Generating Stations
- 494-1974—Method for Identification of Documents Related to Class 1E Equipment and Systems for Nuclear Power Generating System
- 497-198—Trial-Use Criteria for Post Accident Monitoring Instrumentation for Nuclear Power Generating Stations
- 566-1977—Recommended Practice for the Design of Display and Control Facilities for Central Control Rooms of Nuclear Power Generating Stations
- 603-1980—Trial Use Criteria for Safety Systems for Nuclear Power Generating Stations

Lidman, U. 1998. Upgrading of Water and Wastewater Systems. *Water Sci. Tech.* 37(9).

Metcalf & Eddy, Inc., 1991. *Wastewater Engineering: Treatment Disposal and Reuse,* 3rd Ed. New York: McGraw-Hill, Inc.

New England Interstate Environmental Training Center. *Guides for the Design of Municipal Treatment Works (TR-16)*. Portland, ME.

Ontario Ministry of Environment and Energy (OMEE), Water Environment Association of Ontario (WEAO) and Environment Canada (EC). 1996. *Guidance Manual for Sewage Treatment Plant Process Audits*. Ottawa, Ontario, Canada.

Strigini, L. 2000. Fault-Tolerance in Hardware and Software, Course Description, City University, Center for Software Reliability (CSR).

Technology and Design Deficiencies at Publicly Owned Treatment Works. 1989. *Water Environment & Technology*, Alexandria, VA.

U.S. Environmental Protection Agency. 1974. *U.S. EPA Design Criteria for Mechanical, Electrical and Fluid System and Component Reliability*, EPA-430-99-74-001. Washington, DC.

U.S. Environmental Protection Agency. 1978. *General Pretreatment Regulations for Existing and New Sources of Pollution*. 43 FR 27746. Washington, DC.

U.S. Environmental Protection Agency. 1988. *Needs Survey*. Washington, DC.

U.S. Environmental Protection Agency. 1989a. *Handbook: Retrofitting POTWs*, EPA/625/6-89/020. Washington, DC.

U.S. Environmental Protection Agency Office of Municipal Pollution Control, Office of Water. 1989b. *Analysis of Performance Limiting Factors (PLFs) at Small Wastewater Treatment Plants*, EPA WH-546/OMPC 10-89. Washington, DC.

U.S. Environmental Protection Agency. 1990. *General Pretreatment Regulations for Existing and New Sources of Pollution*. 55 FR 30082. Washington, DC.

Water Environment Federation. 1992. *Design of Municipal Wastewater Treatment Plants, Manual of Practice MOP-8*. Alexandria, VA.

Water Environment Federation. 1994. *Pretreatment of Industrial Wastes, Manual of Practice FD-3*. Alexandria, VA.

Water Environment Federation. 1996. *Wastewater Disinfection, Manual of Practice FD-10*. Alexandria, VA.

Water Pollution Control Federation. 1989. *Emergency Planning for Municipal Wastewater Facilities, Manual of Practice SM-8*. Alexandria, VA.

WERF SUSBCRIBERS

WASTEWATER UTILITY

Alabama
Montgomery Water Works & Sanitary Sewer Board

Alaska
Anchorage Water & Wastewater Utility

Arizona
Gila Resources

Glendale, City of, Utilities Department

Mesa, City of

Phoenix Water Services Department

Pima County Wastewater Management

Arkansas
Little Rock Wastewater Utility

California
Calaveras County Water District

Central Contra Costa Sanitary District

Contra Costa Water District

Crestline Sanitation District

Delta Diablo Sanitation District

Dublin San Ramon Services District

East Bay Municipal Utility District

El Dorado Irrigation District

Escondido, City of

Fairfield-Suisun Sewer District

Irvine Ranch Water District

Las Virgenes Municipal Water District

Lodi, City of

Los Angeles, City of

Los Angeles County, Sanitation Districts of

Napa Sanitation District

Orange County Sanitation District

Palo Alto, City of

Riverside, City of

Sacramento Regional County Sanitation District

San Diego Metropolitan Wastewater Department, City of

San Francisco, City & County of

San Jose, City of

Santa Barbara, City of

Santa Rosa, City of

South Bayside System Authority

South Orange County Wastewater Authority

Union Sanitary District

West Valley Sanitation District

Colorado
Boulder, City of

Colorado Springs, City of

Littleton/Englewood Water Pollution Control Plant

Metro Wastewater Reclamation District, Denver

Connecticut
New Haven, City of, WPCA

District of Columbia
District of Columbia Water & Sewer Authority

Florida
Broward, County of

Fort Lauderdale, City of

Gainesville Regional Utilities

JEA

Kissimmee, City of, Department of Water Resources

Miami-Dade Water & Sewer Authority

Orange County Utilities Department

Orlando, City of

Reedy Creek Improvement District

Seminole County Environmental Services

St. Petersburg, City of

Stuart Public Utilities

Tallahassee, City of

Tampa, City of

West Palm Beach, City of

Georgia
Atlanta Department of Watershed Management

Augusta, City of

Clayton County Water Authority

Cobb County Water System

Columbus Water Works

Fulton County

Gwinnett County Department of Public Utilities

Macon Water Authority

Savannah, City of

Hawaii
Honolulu, City and County of

Idaho
Pocatello, City of

Illinois
American Bottoms Wastewater Treatment Plant

Dupage, County of

Greater Peoria Sanitary District

Kankakee River Metropolitan Agency

Metropolitan Water Reclamation District of Greater Chicago

North Shore Sanitary District

Wheaton Sanitary District

Indiana
Fort Wayne, City of

Iowa
Ames, City of

Des Moines Metro Wastewater Reclamation Authority

Kansas
Johnson County Unified Wastewater Districts

Unified Government of Wyandotte County/Kansas City, City of

Kentucky
Louisville & Jefferson County Metropolitan Sewer District

Louisiana
Sewerage & Water Board of New Orleans

Maine
Bangor, City of

Portland Water District

Maryland
Anne Arundel County Bureau of Utility Operations

Howard County Department of Public Works

Washington Suburban Sanitary Commission

Massachusetts
Boston Water & Sewer Commission

Upper Blackstone Water Pollution Abatement District

Michigan
Ann Arbor, City of

Detroit, City of

Grand Rapids, City of

Holland Board of Public Works

Lansing, City of

Owosso Mid-Shiawassee County WWTP

Saginaw, City of

Wayne County Department of Environment

Wyoming, City of

Minnesota
Rochester, City of

Missouri
Independence, City of

Kansas City Missouri Water Services Department

Little Blue Valley Sewer District

Metropolitan St. Louis Sewer District

Nebraska
Lincoln Wastewater System

Nevada
Henderson, City of

New Jersey
Passaic Valley Sewerage Commissioners

New York
New York City Department of Environmental Protection

North Carolina
Cary, Town of

Charlotte/Mecklenburg Utilities

Durham, City of

Metropolitan Sewerage District of Buncombe County

Orange Water & Sewer Authority

Ohio
Akron, City of

Butler County Department of Environmental Services

Columbus, City of

Metropolitan Sewer District of Greater Cincinnati

Northeast Ohio Regional Sewer District

Summit, County of

Oklahoma
Tulsa, City of

Oregon
Clean Water Services

Eugene/Springfield Water Pollution Control

Water Environment Services

Pennsylvania
Philadelphia, City of

University Area Joint Authority, State College

South Carolina
Charleston Commissioners of Public Works

Mount Pleasant Waterworks & Sewer Commission

Spartanburg Sanitary Sewer District

WERF SUBSCRIBERS

Tennessee
Cleveland, City of
Knoxville Utilities Board
Murfreesboro Water & Sewer Department
Nashville Metro Water Services

Texas
Amarillo, City of
Austin, City of
Dallas Water Utilities
Denton, City of
El Paso Water Utilities
Fort Worth, City of
Gulf Coast Waste Disposal Authority
Houston, City of
San Antonio Water System
Trinity River Authority

Utah
Salt Lake City Corporation

Virginia
Alexandria Sanitation Authority
Arlington, County of
Fairfax County Virginia
Hampton Roads Sanitation District
Henrico, County of
Hopewell Regional Wastewater Treatment Facility
Loudoun County Sanitation Authority
Lynchburg Regional WWTP
Prince William County Service Authority
Richmond, City of

Washington
Edmonds, City of
Everett, City of
King County Department of Natural Resources
Seattle Public Utilities
Yakima, City of

Wisconsin
Green Bay Metro Sewerage District
Madison Metropolitan Sewerage District
Milwaukee Metropolitan Sewerage District
Racine, City of
Sheboygan Regional Wastewater Treatment
Wausau Water Works

Australia
South Australian Water Corp.
Sydney Water Corp.
Water Corp. of Western Australia

Canada
Greater Vancouver Regional District
Toronto, City of, Ontario

Mexico
Servicios de Agua y Drenaje de Monterrey, I.P.D.

United Kingdom
Yorkshire Water Services Limited

STORMWATER UTLITY

California
Los Angeles, City of, Department of Public Works
Monterey, City of
Santa Rosa, City of

Colorado
Boulder, City of

Georgia
Griffin, City of

Kansas
Overland Park, City of

Kentucky
Louisville & Jefferson County Metropolitan Sewer District

Maine
Portland Water District

North Carolina
Cary, Town of
Charlotte, City of, Stormwater Services

Pennsylvania
Philadelphia, City of

STATE

Fresno Metropolitan Flood Control District, Calif.
Urban Drainage & Flood Control District, Colo.

CORPORATE

ADS Environmental Services
The ADVENT Group Inc.
Alan Plummer & Associates
Alden Research Laboratory
Alpine Technology Inc.
Aquateam–Norwegian Water Technology Centre A/S
BaySaver Inc.
BioVir Inc.
Black & Veatch
Boyle Engineering Corporation
BPR CSO
Brown & Caldwell
Burns & McDonnell
CABE Associates Inc.
The Cadmus Group
Camp Dresser & McKee Inc.
Carollo Engineers Inc.
Carpenter Environmental Associates Inc.
CDS Technologies Inc.
Chemtrac Systems Inc.
CH2M HILL
Clancy Environmental Consultants Inc.
Damon S. Williams Associates, LLC
David L. Sheridan, P.C.
Earth Tech Inc.
Ecolab Water Care Services
EMA Inc.
The Eshelman Company Inc.
Finkbeiner, Pettis, & Strout Inc.
Frontier Geosciences Inc.
ftn Associates Inc.
Gannett Fleming Inc.
GE Betz
Golder Associates Inc.
Greeley and Hansen LLC
The HACH Company
Hazen & Sawyer, P.C.
HDR Engineering Inc.
HNTB Corporation
HydroQual Inc.
Insituform Technologies Inc.
Institute for Environmental Technology & Industry, Korea
Jacobson Helgoth Consultants Inc.
Jason Consultants Inc.
Jordan, Jones, & Goulding Inc.
KCI Technologies Inc.
Kelly & Weaver, P.C.
Kennedy/Jenks Consultants
Komline Sanderson Engineering Corporation
Lawler, Matusky & Skelly Engineers, LLP
Limno-Tech Inc.
Lombardo Associates Inc.
Malcolm Pirnie Inc.
McKim & Creed
MEC Analytical Systems Inc.
MWH
New England Organics
Odor & Corrosion Technology Consultants Inc. (OCTC)
ONDEO Degremont Inc.
Oswald Green, LLC
PA Government Services Inc.
Parametrix Inc.
Parsons
Post, Buckley, Schuh & Jernigan
R&D Engineering/Conestoga Rover & Associates
The RETEC Group
R.M. Towill Corporation
Ross & Associates Ltd.
Royce Technologies
SAIC Maritime Technical Group
Sear-Brown
Sierra Environmental Services
SYNAGRO
Tetra Tech Inc.
Trojan Technologies Inc.
URS Corporation
USFilter NATC
Vortechnics Inc.
Wade-Trim Inc.
Weston Solutions Inc.
Woodard & Curran
WRc/D&B, LLC
WWETCO, LLC

INDUSTRY

American Electric Power
ChevronTexaco Energy Research & Technology Company
The Coca-Cola Company
Dow Chemical Company
DuPont Company
Eastman Kodak Company
Merck & Company Inc.
ONDEO Services
Procter & Gamble Company
PSEG Services Corp.
Severn Trent Services Inc.
Shell Global Solutions
Thames Water Plc
United Water Services LLC

Note: List as of 5/12/03

WERF Product Order Form

As a benefit of joining the Water Environment Research Foundation, subscribers are entitled to receive one complimentary copy of all final reports and other products. Additional copies are available at cost (usually $10). To order your complimentary copy of a report, please write "free" in the unit price column. WERF keeps track of all orders. If the charge differs from what is shown here, we will call to confirm the total before processing.

Name ______________________ Title ______________________

Organization ______________________

Address ______________________

City ______________ State ________ Zip Code ________ Country ________

Phone ______________ Fax ______________ Email ______________

Stock #	Product	Quantity	Unit Price	Total
			Postage & Handling	
			VA Residents Add 4.5% Sales Tax	
			Canadian Residents Add 7% GST	
			TOTAL	

Method of Payment: (All orders must be prepaid.)

❑ Check or Money Order Enclosed

❑ Visa ❑ Mastercard ❑ American Express

Account No. ______________________ Exp. Date ________

Signature ______________________

Shipping & Handling:

Amount of Order	United States	Canada & Mexico	All Others
Up to but not more than:	Add:	Add:	Add:
$20.00	$5.00*	$8.00	50% of amount
30.00	5.50	8.00	40% of amount
40.00	6.00	8.00	
50.00	6.50	14.00	
60.00	7.00	14.00	
80.00	8.00	14.00	
100.00	10.00	21.00	
150.00	12.50	28.00	
200.00	15.00	35.00	
More than $200.00	Add 20% of order	Add 20% of order	

*minimum amount for all orders

Note: Please make checks payable to the Water Environment Research Foundation.

To Order (Subscribers Only):

Log on to www.werf.org and click on the "Product Catalog."

Phone: (703) 684-2470
Fax: (703) 299-0742.

WERF
Attn: Subscriber Services
601 Wythe Street
Alexandria, VA 22314-1994

To Order (Non-Subscribers):

Non-subscribers may be able to order WERF publications either through WEF (www.wef.org) or IWAP (www.iwapublishing.com). Visit WERF's website at www.werf.org for details.